Disaster Resilience in South Asia

This book considers the challenges of building disaster resilience in South Asia – a region that frequently experiences some of the most severe and devastating impacts of disasters.

Despite significant work to assist affected communities, many smaller South Asian countries remain particularly vulnerable in terms of fostering disaster resilience. Drawing on examples from Bangladesh, Bhutan, Maldives, Nepal and Sri Lanka, the book offers rich insights and narratives on disaster resilience policy and practice. It considers the possibilities for advancing community resilience and capacity building through an exploration of different aspects of governance and policy. Given the diversity of these countries and recent disasters, a variety of perspectives are considered: institutional and policy frameworks, risk management governance, recovery operations, building codes, and policy and media discourse. The book offers a collective understanding of practice, which can offer global lessons to a world increasingly beset by disasters and with uncertain environmental futures.

This book will be a valuable resource for scholars, practitioners and students in the fields of disaster risk reduction and management, climate change adaptation, public policy and sustainable development.

Iftekhar Ahmed is Associate Professor at the University of Newcastle, Australia. His research focuses on disaster resilience, post-disaster housing and climate change adaptation. He has co-authored the book *Sustainable Housing Reconstruction* (Routledge, 2015) and co-edited the book *Community Engagement in Post-Disaster Recovery* (Routledge, 2017).

Kim Maund is Discipline Head and Senior Lecturer in Construction Management at the University of Newcastle, Australia. With more than 15 years' industry experience across government and private sectors, her research focuses on policy within the built environment, particularly in relation to urban planning and building and codes in the context of disaster resilience.

Thayaparan Gajendran is Associate Professor at the University of Newcastle, Australia. His research focuses on cultural and governance approaches for building resilience. He has co-authored many research publications in the areas of disaster risk communication and disaster governance.

Routledge Studies in Hazards, Disaster Risk and Climate Change

Series Editor: Ilan Kelman

Reader in Risk, Resilience and Global Health at the Institute for Risk and Disaster Reduction (IRDR) and the Institute for Global Health (IGH), University College London (UCL)

This series provides a forum for original and vibrant research. It offers contributions from each of these communities as well as innovative titles that examine the links between hazards, disasters and climate change, to bring these schools of thought closer together. This series promotes interdisciplinary scholarly work that is empirically and theoretically informed, with titles reflecting the wealth of research being undertaken in these diverse and exciting fields.

Disasters, Vulnerability, and Narratives
Writing Haiti's Futures
Kasia Mika

Climate Change Impacts and Women's Livelihood
Vulnerability in Developing Countries
Salim Momtaz and Muhammad Asaduzzaman

Risk Communication and Community Resilience
Edited by Bandana Kar and David M. Cochran

Climate Hazards, Disasters, and Gendered Ramifications
Catarina Kinnvall and Helle Rydström

The Anthropology of Disasters in Latin America
State of the Art
Edited by Virginia García-Acosta

Disaster Resilience in South Asia
Tackling the Odds in the Sub-Continental Fringes
Iftekhar Ahmed, Kim Maund and Thayaparan Gajendran

For more information about this series, please visit: www.routledge.com/Routledge-Studies-in-Hazards-Disaster-Risk-and-Climate-Change/book-series/HDC

Disaster Resilience in South Asia

Tackling the Odds in the Sub-Continental Fringes

Iftekhar Ahmed, Kim Maund and Thayaparan Gajendran

Routledge
Taylor & Francis Group
LONDON AND NEW YORK

First published 2020
by Routledge
2 Park Square, Milton Park, Abingdon, Oxon OX14 4RN

and by Routledge
605 Third Avenue, New York, NY 10017

First issued in paperback 2021

Routledge is an imprint of the Taylor & Francis Group, an informa business

Publisher's Note
The publisher has gone to great lengths to ensure the quality of this reprint but points out that some imperfections in the original copies may be apparent.

British Library Cataloguing-in-Publication Data
A catalogue record for this book is available from the British Library

Library of Congress Cataloging-in-Publication Data
A catalog record has been requested for this book

ISBN 13: 978-1-03-223636-0 (pbk)
ISBN 13: 978-1-138-37002-9 (hbk)

DOI: 10.4324/9780429428296

Typeset in Times New Roman
by Swales & Willis, Exeter, Devon, UK

Contents

Figures

Tables

Acknowledgements

Numerous people around the world have supported, encouraged and inspired us to write this book. Many of them in the South Asian countries featured in the book deal with the challenges of disasters on a regular basis and we are indebted to them for so graciously sharing their knowledge and experience.

Acknowledgements are extended to Arif Abdullah Khan, Programme Specialist, Disaster Resilience, UNDP Bangladesh, and Kamal Hossain, Project Manager, National Resilience Programme, DDM, MoDMR and UNDP Bangladesh, for sharing information, publications and other resource materials, and also for facilitating the fieldwork in Bangladesh and providing connections to experts and other persons who are knowledgeable about CDMP and disaster resilience in Bangladesh. Also, acknowledgement to UNDP Bangladesh for commissioning the consultancy on 'Feasibility Study and Proposal Writing for Strengthening National Adaptive Capacity through Climate Resilient Rural Housing in Coastal Bangladesh', from which material for the Bainpara project was derived.

Acknowledgements are due to Ugyen Dorji, Climate Change Policy Specialist, UNDP Bhutan, for his advice and support for the fieldwork in Bhutan, and also for sharing information, publications and other resource materials. Acknowledgements also to Sonam Deki, Program Officer, Department of Disaster Management, Bhutan, and Karma Dupchu, Director, National Centre of Hydrology and Meteorology, Bhutan, and his colleagues for sharing their knowledge and information. Acknowledgements are also due to Bishnu Chettri, Intern, UNDP Bhutan, and Singhbir Tamang for being interpreters and supporting the fieldwork in Bhutan in different ways.

Acknowledgements are due to Umar Fikry, Manager of Programs and Services, Maldivian Red Crescent, and Rishana Fathimath for organising the fieldwork in the Maldives, and to Hisan Hassan, Project Director, NDMC, for sharing information and other resources on the Dhuvaafaru project in the Maldives. Acknowledgements also to Nooman Ibrahim and Mohamed Izman Suhail for being interpreters and supporting the Maldives fieldwork in various ways. An acknowledgement is also extended to Leeanne Marshall, Shelter Technical Lead, Australian Red Cross, Melbourne, Australia, for facilitating the connection with the Maldivian Red Crescent.

Acknowledgements are extended to Hari Darshan Shrestha, Nagendra Raj Sitoula and Inu Pradhan-Salike of Tribhuvan University, Nepal, for providing the draft country report for our collaborative research project and also for other information that was drawn upon in this chapter. Acknowledgements are also due to Jerry Chati Tasantab, PhD scholar, University of Newcastle, Australia, for sourcing data for the chapter on Nepal. The Asia Pacific Network for Global Change Research (APN), Japan, is acknowledged for supporting the research project on 'Understanding the Opportunities and Challenges of Compliance to Safe Building Codes for Disaster Resilience in South Asia – the Cases of Bangladesh and Nepal' which informed the Nepal chapter.

The authors wish to acknowledge and thank their project partners in Sri Lanka for their dedication and support: Dr Yasangika Sandanayake, Dr Thilini Jayawickrama and Mr Tharindu Karunaratne from the University of Moratuwa, Sri Lanka. Acknowledgements are due to the following organisations who contributed to case study activities: Asian Disaster Preparedness Centre, Department of Meteorology Sri Lanka, Disaster Management Centre Sri Lanka, Divisional Secretariat Ratnapura, Sri Lanka, Federation of Sri Lanka Local Government Authorities, Irrigation Department – Ratnapura, Sri Lanka, Kothalawala Defence University, Sri Lanka, National Building Research Organisation, Sri Lanka, National Disaster Relief Services Centre, Sri Lanka, United Nations Development Programme and the Urban Development Authority – Ratnapura, Sri Lanka. The Faculty of Engineering and Built Environment, University of Newcastle, Australia, is acknowledged for providing a Strategic Pilot Grant for the research project 'Strengthening the Research Base in South Asia: Disaster Risk Reduction in the Context of Recent Disasters in Sri Lanka' that informed the Sri Lanka chapter.

Some of the key contributors who we gratefully thank for helping the production of the book include: Bastian Fox Phelan for highly insightful and incisive editing; Rumana Asad for producing the maps with diligence and patience; Ekushey Ahmed and Shumank Deep for the tedious task of putting together and checking all the references. And, importantly, Ben Wisner for writing such an insightful and thought-provoking foreword.

Finally, we would not have been able to write this book without the support and enthusiasm, as well as patience, of our respective families.

Foreword

Ben Wisner

In thinking about disaster risk in Asia, scholars and planners too often fall prey to easy stereotypes without digging deeply into the lived reality of these statements. 'Most hazard-prone region of the world' is repeated in study after study. The phrase has become empty of its power to shock, to focus attention and to communicate both the urgency of action and the respect due to citizens, scientists, citizen-scientists and leaders living and even thriving against these odds. I congratulate the authors of this book for breathing life back into the phrase in a fresh and important contribution. They treat five smaller countries in South Asia in very substantial, well-illustrated, clearly written chapters, each of which is close to the physical and socio-political realities of these smaller countries and produced with the acknowledged assistance of local experts.

The authors provide counterpoint and corrective to a view that would give too much dominance to the approaches to disaster risk management (DRM) developed in the region's largest country, India. Although Maldives and Nepal have learned from the approaches taken in some Indian states, and also have benefitted from technical and financial assistance from their large neighbour, this book shows that the smaller countries have independently developed their own unique institutions and approaches. International cooperation with UN agencies, the EU and bilateral donors, and international non-governmental organizations (INGOs) is shown to have had an important role in tackling such major challenges as glacial lake outburst floods (GLOFs) in Bhutan, a massive rebuilding effort in Nepal following the 2015 Gorkha Earthquake, Indian Ocean Tsunami damage in Maldives, Sri Lanka's 2017 floods and landslides and been key to the strides made in DRM in Bangladesh.

This has not been easy, and the authors do not shy away from the continuing problems of potential dependency on external support, the danger that projects will not be fully owned and internalised into daily lives (say, in the example of early warning systems in GLOF-exposed valleys or building codes among the thousands of local masons and small contractors in Nepal) and in governments' institutions and recurrent budgets. Thus, *Disaster Resilience in South Asia* is also a corrective to triumphalism and points to the continuing need for education and capacity-building, improved institutional coordination, better integration of formal and informal DRM practices (where local consultation is vital). Both the introductory and concluding chapter provide a realistic view of the gaps.

This timely and important book also helps to rescue the concept of 'resilience'. Over the past decade, this word has become a portmanteau that sometimes obscures more than it illuminates. 'Resilience' has been misinterpreted by some as an invocation of individual responsibility to secure one's own life and well-being in an age of austerity and precarious employment with less assistance from the state. This is a blatantly neo-liberal reading that re-purposes the term's original use in materials science and ecology as a justification for privatisation (Klein, 2008). By contrast, the authors take the pains to define how they use the word very clearly and resort to more useful definitions. They cite and quote from four contemporary definitions of 'resilience'. The first is that of the United Nations Office for Disaster Risk Reduction (2019). It is the least ambitious, going only as far as 'restoration of essential basic functions'. This is problematic in the five countries considered, where, as the authors state (pp. 1–2), citing the United Nations Economic and Social Commission for Asia and the Pacific (2019), that in these countries hazards such as drought and flood are superimposed on considerable pre-existing, baseline 'poverty, hunger and undernourishment'. However, the authors also rely on definitions from UN-Habitat (2019) and the UK's overseas development department (Department for International Development, 2011). These definitions stress 'transformation' and sustainability (although the latter term is also a very slippery one, especially when applied to baseline conditions of inequality and exclusion!). Finally, they choose a strong definition from within the region developed by the UN Economic and Social Commission for Asia and the Pacific (UNESCAP, 2019). This one makes clear that resilience involves establishing conditions in which major hazard events do not 'disempower vulnerable groups', 'aggravate poverty' and cause a 'widening of inequality'.

What about inequality? Three of the chosen countries have seen a decrease in inequality from 1980 to 2015 (Maldives, Bhutan and Nepal); whilst Bangladesh and Sri Lanka have experienced an increase (Salim, 2019). One wonders how Maldives and Bhutan, and especially Nepal, have been able to achieve this shift in the direction of more equal distribution of income in the face of frequent hazard events and the impacts of climate change that are arguably as severe and erosive of livelihood security as those in Bangladesh and Sri Lanka, where inequality has increased?

This book is rich in detail. Across the five core chapters, multiple-lenses provide a necessary reality-check. The reader learns how long recovery takes and appreciates the magnitude of the large hazards faced even though small and medium sized hazard events are faced more frequently, and their aggregate erosive effect on livelihood and well-being (even happiness, as measured annually since 2018 in Bhutan) is very large. One also learns once more how much admiration and respect is due to leaders and citizens in these countries who are living against these odds.

References

Department for International Development (DFID). (2011) *Defining Disaster Resilience: A DFID Approach Paper*. London: DFID.

Klein, N. (2008) *The Shock Doctrine: The Rise of Disaster Capitalism*. New York: Picador.

Salim, R. (2019) *Cross-country Differences in Income Inequality – Where Do South Asian Countries Stand?* Available at: www.ipsnews.net/2016/09/cross-country-differences-in-income-inequality-where-do-south-asian-countries-stand/ (accessed 7 November 2019).

UN-Habitat. (2019) *Resilience*. Available at: https://unhabitat.org/resilience/ (accessed 5 March 2020).

United Nations Office for Disaster Risk Reduction (UNDRR). (2019) *Terminology on Disaster Reduction*. Geneva: UNDRR.

United Nations Economic and Social Commission for Asia and the Pacific (UNESCAP). (2019) *Asia-Pacific Disaster Report 2019: The Disaster Riskscape across Asia-Pacific, Pathways for Resilience, Inclusion and Empowerment*. Bangkok: UNESCAP.

1 Introduction

Tackling the odds in the sub-continental fringes: disaster resilience in the smaller countries of South Asia

Overview

The world is continuously and widely affected by hazards of various kinds, resulting in large catastrophic disasters such as tsunamis, earthquakes and cyclones with massively damaging impacts, to regularly occurring disasters such as floods, drought and wildfires that insidiously gnaw away assets and development gains. Global warming and the impacts of climate change contribute to creating an even more complex and often uncertain future with regard to disasters (United Nations Economic and Social Commission for the Asia Pacific, 2019). Globally, in the last couple of decades more than 1.35 million human lives were lost to disasters with almost 70 per-cent attributed to lower income countries (Centre for Research on the Epidemiology of Disasters and United Nations Office for Disaster Risk Reduction, 2015). The economic cost of disasters continues to grow, amounting to more than US$350 billion in 2017 (United Nations Office for Disaster Risk Reduction, 2019a).

Asia is a region subjected to frequent and devastating natural hazards (UNESCAP, 2019). Considered the most severely and extensively affected continent throughout recent history in terms of number of disasters, people killed or affected and economic damage (Asian Disaster Reduction Center, 2016; Guha-Sapir et al., 2016): potentially due to the large concentration of the world's population accounting for more than 60 per-cent (United Nations, 2018). UNESCAP (2019) reported that of the 281 global disasters recorded in 2018 approximately half occurred across the Asian Pacific region. Furthermore, 'eight of the ten most harmful natural disasters transpired within the region' (UNESCAP, 2019, p. 6).

Two of the most populous countries in the world, China and India, are in Asia. These two countries are also the top two emerging economies across the region with recent remarkable increases in wealth, physical assets and infrastructure (Leading Edge Alliance, 2017) and consequently human development including capacity for achieving disaster resilience in terms of addressing disaster risks and recovering from disasters. However, for many other countries within the Asian Pacific region, vulnerability to disasters remains high and the ability to develop resilience limited. UNESCAP (2019) identify that unfortunately those most vulnerable to disasters 'are often in remote areas, or who belong to minority groups or live in the fragile margins of big cities, with little defence against disasters' (p. 106).

Many of the smaller Asian countries are in such a position where disaster impacts are often more severe, human suffering deeper, and unfortunately, institutional and community capacity for disaster resilience is usually weaker. South Asia is often the most disaster-impacted region: in 2015 almost 65 per cent of the global disaster fatalities occurred across this area (UNESCAP, 2015). Furthermore, UNESCAP (2019) has identified many countries in the region that experience disasters such as floods and droughts while being in a position of severe 'poverty, hunger and undernourishment' (p. 20). Moreover, transboundary river basins are often regions that shelter the poor and, subsequently, those who rely upon local agricultural practice (UNESCAP, 2019) and live on land that is considered at high risk of disaster events.

A number of smaller countries belong to the Asian region including Bangladesh, Bhutan, the Maldives, Nepal and Sri Lanka (Compare Infobase Limited, 2012), all of which are threatened by various natural hazards and are economically vulnerable. Despite significant work to assist affected communities, many smaller South Asian countries are not in a position of economic wealth; resilience is influenced by, *inter alia*, poverty, ineffective governance, urbanisation, limited resources and environmental degradation. Disaster events substantially erode community assets, as these smaller regions struggle to preserve assets and invest in vital infrastructure (United Nations International Strategy for Disaster Reduction, 2015). Ultimately, the disaster impacts in these countries are often most severe, as the ability to develop adequate disaster resilience remains limited while vulnerability reigns. Nonetheless, despite the constraints there are notable efforts in the region to achieve resilience, some of which are highlighted in this book.

In light of the abovementioned, as well as the need to further understand the complexities around vulnerability and enhancing resilience, this book has a focus upon five South Asian countries: Bangladesh, Bhutan, Maldives, Nepal and Sri Lanka (refer Figure 1.1).

Bangladesh

The riverine country of Bangladesh is comprised of many of low-lying plains and river deltas. Bounded predominantly by India and to some extent Myanmar, the south is bounded by the Bay of Bengal – this exposes the country to the full force of hazard events such as cyclones and storm surges with major riverine flooding common during monsoon seasons. There is a strong risk of earthquakes with high vulnerability in the country's rapidly growing and densely populated cities. Bangladesh is acknowledged as a country subject to devastating and ongoing disaster events. From a government perspective, the development of disaster resilience initiatives over recent decades has had a significant impact upon reducing death tolls from the hundreds of thousands who sadly

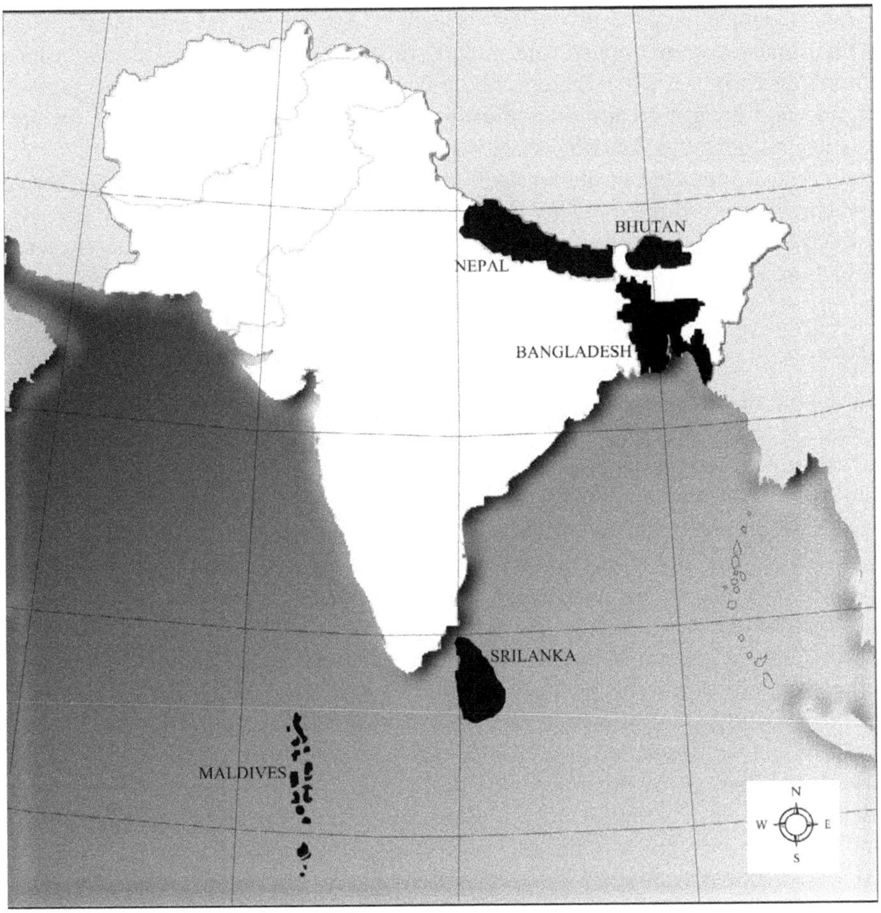

Figure 1.1 Map of South Asia showing the case study countries of this book

perished in devastating events during the 1970 to 1990 period. However, loss of life continues to occur, and devastation in terms of property, infrastructure, business and national economic impact continues to affect millions nationally. Thus, a high degree of vulnerability continues to exist as local communities attempt to reconstruct their homes and livelihoods.

Bhutan

Bhutan, nestled between China and India, is a land-locked country with no direct coastal connection. However, given its proximity to Nepal on the east, it is surrounded by the Himalayas and presents as a country full of mountainous slopes and glacial lakes. These lakes are of primary concern and present a rather unique hazard, given the propensity for glacial lake outburst floods. With climate change impacts and melting of glaciers, the ongoing threat to local communities is a reality. Importantly, the glacial melts enable existing lakes to expand and merge, creating significant water bodies with the potential for devastating outburst floods. Bhutan is a rather closed country in that it maintains traditional culture and limits tourism. From a disaster risk resilience perspective, the country has limited resources to reduce vulnerability and build capacity. Initiatives have been implemented, yet with limited expertise in disaster risk reduction, assistance is often sought externally.

Maldives

Surrounded by the Indian Ocean, the small low-lying coral islands of the Maldives are often afflicted by hazards – particularly of a hydro-meteorological nature, much like their neighbours. Frequent windstorms, low intensity cyclones and thunderstorms involving heavy rainfall, as well as sea surges have been experienced across the region; however, these have largely occurred around the northern atolls. Conversely, the southern islands are considered at risk of earthquakes, given their proximity to local fault lines: mostly, these have been associated with mild tremors. The atolls are approximately 1–1.5 metres above sea level; therefore, global warming-induced climate change effects present the most significant future risk to local communities, particularly given sea level rise. In many respects, the Maldives may be considered a country afflicted by low-intensity events. However, the 2004 Indian Ocean tsunami presented a very different situation, given the force at which the sea waves attacked. The resulting impacts were devastating and exposed the region in terms of its true vulnerability and lack of preparedness to hazards and subsequent disaster events.

Nepal

Nepal is located in the central part of the Himalayas and is bordered by China and India. The country contains eight of the world's highest mountains including Mount Everest. Nepal is affected by frequent hazards that range from earthquakes, severe thunderstorms and flooding through to extreme heat and drought. Environmental issues such as acidification and deforestation plague many regions and often exacerbate the impacts of hazards. Most notably, Nepal is located at the boundary of the actively converging Indo-Australian plate and Eurasian Plate, rendering the country one of the most earthquake-prone areas. The area is also within the seismic zone between India and the Eurasian plates; therefore, the risk and magnitude of frequent earthquakes exposes communities to significant devastation. The 2015 Gorkha earthquake – considered the worst in over 80 years – contributed to a significant loss of life. Tens of thousands were injured, with hundreds of thousands of buildings and infrastructure services damaged. With many community members residing in remote and rural areas, the ability for rescue operations and support presented an extremely difficult situation. Events, such as the 2015 earthquakes highlight an urgent need to build capacity on a local level and increase preparedness, given the likelihood of future seismic events.

Sri Lanka

Sri Lanka, an island separated from the Indian peninsula by the Palk Strait, remains isolated from the mainland and is frequently affected by hazard events of a hydro-meteorological nature. Due to its isolation, it experiences disasters from flooding to landslides with devastating consequences. The death toll and reported missing persons resulting from these frequent events truly reflects a high level of vulnerability experienced on a national level. Furthermore, damage to property and land is often significant. Essential services including health, education and transportation are frequently affected, presenting challenges to local communities and consequently to the national economy. The Government has established a range of agencies and policies to assist with disaster risk reduction; however, the overwhelming loss of life and physical damage reflects a need to build further capacity and resilience. The devastation caused by the 2017 flooding and landslide disasters sadly reflects the scale of vulnerability on a local level, whereby local communities to this day continue to rebuild their lives, all the while exposed to the next disaster event.

As depicted above, these five countries experience frequent and devastating hazards, many of a hydro-meteorological nature and their subsequent disaster events. Regardless of land mass size or population numbers, history shows that there remains a high degree of vulnerability. With predictions of ongoing hazard exposure and climate change, this presents both a concerning and uncertain future. Within this context, there are disaster resilience initiatives that are remarkable given the severe challenges that prevail. The question is then: how can vulnerability be reduced amongst communities, given future hazard predictions. Research, particularly from a government context, across these smaller nations is often limited; yet there is increasing attention towards the need for understanding key influences through critical analysis to inform evidence-based disaster resilience.

The term 'resilience' is now widely used in different discourses, with different intents and has become somewhat a tool for advancing human well-being and dignity. In its base form, the term has been defined as 'The capacity to recover quickly from difficulties' (Oxford Dictionary, 2019). Applied in the content of urban communities, UN-Habitat provide an interpretation that promotes adaptation for improved liveability:

> the ability of any urban system to maintain continuity through all shocks and stresses while positively adapting and transforming towards sustainability. Therefore, a resilient city is one that assesses, plans and acts to prepare for and respond to all hazards, either sudden or slow-onset, expected or unexpected. By doing so, cities are better able to protect and enhance people's

> lives, secure development gains, foster an investible environment and drive positive change.
>
> (UN-Habitat, 2019)

While 'resilience' is a broad concept and is interpreted according to various agendas and persuasions relating to a range of issues, this book focuses specifically on 'disaster resilience', drawing on concepts presented by organisations such as the Governance and Social Development Resource Centre (Combaz, 2014) and United Nations Office for Disaster Risk Reduction (UNDRR, 2019b). In this context, UNDRR defines resilience as:

> the ability of a system, community or society exposed to hazards to resist, absorb, accommodate to and recover from the effects of a hazard in a timely and efficient manner, including through the preservation and restoration of its essential basic structures and functions.
>
> (UNDRR, 2019b)

How can resilience be approached in this setting, especially given that 'A rapid sequence of disasters can outpace people's resilience, eroding development gains by aggravating poverty, intensifying environmental degradation, disempowering vulnerable groups and widening inequalities' (UNESCAP, 2019, p. 23). The UK Department for International Development (DFID) defines resilience in a similar way to the definition presented above; however, they identified a framework consisting of four elements that identified 'core elements' in the context of disaster resilience. DFID explains that:

> Disaster Resilience is the ability of countries, communities and households to manage change, by maintaining or transforming living standards in the face of shocks or stresses – such as earthquakes, drought or violent conflict – without compromising their long-term prospect.
>
> (DFID, 2011, p. 6)

In this respect, the framework consists of the following: context; disturbance; capacity to respond; and reaction. Context refers to the assemblage or system for which resilience is aimed, such as a particular group or institution. Disturbance then progresses into the type of activities and events resilience is aimed towards, such as disasters and climate change. Capacity to respond quite simply refers to the ability to cope with a shock or stress, as related to exposure or magnitude; sensitivity or degree affected; and adaptive capability or ability to adjust. While reaction refers to response actions: defined as bounce back better, bounce back or recover. Respectively, these refer to the ability to absorb stress and shock and use this to build greater resilience; the ability to bring back conditions that were pre-existing; or the reduced capacity and ability to cope (DFID, 2011).

Although an understanding of disaster resilience exists, from a research perspective there are not many books on this area focusing on the South Asian region. Most of the existing books consider themes around gender, disaster preparedness and disaster response in large countries. Moreover, they provide an overview of disaster resilience initiatives and project evaluations that are generally presented in a reporting format, primarily as grey literature. Therefore, this book is unique, given its capacity to explore disaster resilience initiatives. To understand the possibilities for advancing community resilience and capacity building through an exploration of different aspects of governance, policy and action, this book focuses on five case study countries: Bangladesh, Bhutan, Maldives, Nepal and Sri Lanka. Given the diversity of the regions and recent disasters, each study takes an individual approach of employing a specific lens to provide a collective understanding of practice: institutional and policy frameworks; risk management governance; recovery options; building codes; and media discourse.

Chapter overviews

Bangladesh

Bangladesh – a country that has been dealing with frequent hazards for many years – is presented as 'An old hand': exploring the institutional relationships to disaster risk reduction between the government, United Nations and non-governmental organisations (NGOs). Bangladesh has been known for a long time for its disasters, but it is also one of the first countries to have established a dedicated ministry dealing with disasters. Bangladesh is a context with a plethora of NGOs and a strong presence of UN agencies. The research examines the institutional and policy framework for disaster resilience, within which a key sector such as housing is explored. It thus offers lessons on a range of institutional relationships, approaches, synergies and oppositions.

Bhutan

The uniqueness of flooding in Bhutan evokes a discussion around the 'Icy flood': adapting to the impacts of glacial lake outburst floods. A unique hazard, yet understanding risk reduction measures offers wider lessons for a climate change-induced melting planet. How disaster resilience policy and actions are implemented in a monarchy within a highly isolationist country brings a unique narrative to the disaster resilience discourse. Yet it allows for questioning of the persuasive rhetoric of more established and widespread disaster governance systems elsewhere.

Maldives

The Maldives are depicted as a country 'Staying afloat': the focus is on understanding long-term disaster recovery in the housing sector in the climate change-threatened atolls. Following the 2004 Indian Ocean tsunami, significant housing reconstruction projects were implemented across the Maldives, yet research has been comparatively scanty. Recovery of the housing sector is often a protracted process with its wider linkages. Therefore, long-term outcomes in this sector after a decade indicate prospects for adaptation, especially in a context so severely threatened by climate change, and the emergence of governance mechanisms to deal with the evolving challenges since the tsunami is considered. The intent of this chapter is to explore disaster resilience in terms of recovery activities following the 2004 tsunami.

Nepal

Nepal presents a 'Rule of thumb' option: looking at opportunities and barriers to compliance with building codes for disaster resilience in relation to earthquakes. Earthquakes pose a serious risk in Nepal and disaster-resilient building codes are essential to reduce impacts, an issue that has come into sharp focus after the massive earthquake event in 2015. Recovery from the earthquake has been slow, but on the other hand, the government and other stakeholders have been taking the implementation of building codes seriously. Although Nepal is far from experiencing widespread compliance, it is a small step towards achieving greater disaster resilience.

Sri Lanka

Within the context of Sri Lanka, the theme is 'Framing accountability': examining the policy and media discourse on disaster risk reduction and management. While the devastation by the Indian Ocean tsunami in 2004 was massive and widespread, Sri Lanka continues to experience more localised hazards such as floods and landslides on a regular basis, such as the devastating events during 2017. Media and policy discourse following this event reveals the achievements and shortcomings in disaster risk reduction and management approaches, and points to the way forward for long-term disaster resilience.

All these case studies provide narratives on tackling immense odds for achieving disaster resilience. With widespread human development, these countries often face significant challenges: poverty; inadequate governance; unplanned urbanisation; endemic corruption; pressure on limited resources; inequitable power and resources distribution; and environmental stress from land degradation, over-exploitation of natural assets and pollution. These countries are also highly vulnerable to a range of natural hazards, including: floods of different types including monsoonal, flash, tidal, glacial; drought; windstorms such as cyclones and tornadoes; earthquakes, tsunamis and landslides; as well as human-induced hazards such as water-logging, fire, groundwater contamination, industrial accidents and building collapse. A range of climate change impacts are also evident, including sea level rise, salinity intrusion, and heat and cold stress. Most of the engagements in this book have the common underlying aim of understanding possibilities for advancing disaster resilience – this is undertaken through an exploration of a range of issues pertaining to this aim.

These case studies comprise a rich repository of knowledge from the smaller South Asian countries on different aspects of disaster resilience: all the more valuable because of the efforts there for cultivating resilience despite the conditions of extreme constraints. While the case studies are discrete, compiling them into the form of a book offers a rich insight into disaster resilience policy and practice across the region. The book illustrates a wider set of lessons in terms of common experiences and features unique and remarkable contextual vignettes. Thus, in this manner, it helps advance knowledge and understanding of the South Asian region with its vast agglomerations of people and cultures, its context of devastating disasters, and its prospects for capacity building and achieving resilience. The diversity of the settings and the array of disaster resilience efforts offer wider global lessons to a world increasingly beset by disasters and facing an uncertain future of climate change impacts.

There is increasing attention placed on understanding disaster resilience through 'critical analysis' employing 'multiple lenses'. The critical framing of the 'context' is key to inform evidence-based disaster resilience discussed and practised among academic and practitioner communities. Therefore, a book providing a solid discourse on contextual factors with critical analysis of key issues influencing the disaster resilience is indispensable for advancement of this field.

This book is unique in two ways:

1. The collection of the nations in terms of disaster resilience: a true combination of the smaller South Asian counties focusing on disaster resilience.
2. Use of multiple lenses to study disaster resilience: unique methodologies to explore resilience issues in each country.

The research discussed within this book is informed by social lenses, particularly through engagement with analysis at the social-structural level in understanding disaster resilience in the case study countries. Seeking options for cultivating resilience within these smaller and most vulnerable communities is considered as key influences are highlighted to reinforce policy intent, capacity building and the achievement of disaster resilience on a regional level. Importantly, concepts of disaster resilience in the book are derived from local stakeholders, including institutional actors and at-risk/affected communities, combined with learning from the ground reality by the authors in a constructivist paradigm to accommodate multiple interpretations of reality.

The devastating and high profile disasters mentioned above that occur across the smaller nations of South Asia continue to raise awareness of the importance of building capacity and reducing vulnerability. They have spurred significant institutional and community interest and activity in these countries on disaster resilience initiatives. Research endeavours such as those identified within this book assist to achieve these outcomes through knowledge dissemination and identifying challenges as well as barriers and enablers to change.

References

Asian Disaster Reduction Center (ADRC). (2016) *Natural Disaster Data Book 2015: An Analytical Overview*. Kobe: ADRC.

Centre for Research on the Epidemiology of Disasters (CRED) and United Nations Office for Disaster Risk Reduction (UNDRR). (2015) *Poverty and Death: Disaster Mortality 1996–2015*. Belgium/Switzerland: CRED/UNISDR.

Combaz, E. (2014) *Disaster Resilience: Topic Guide*. Birmingham: Governance and Social Development Resource Centre (GSDRC), University of Birmingham.

Compare Infobase Limited. (2012) *SAARC Country Map*. Maps of India. Available at: www.mapsofindia.com/maps-of-asia/saarc-country-map.html (accessed 7 March 2018).

Department for International Development (DFID). (2011) *Defining Disaster Resilience: A DFID Approach Paper*. UK: DFID.

Guha-Sapir, D., Hoyois, P., Wallemacq, P. and Below, R. (2016) *Annual Disaster Statistical Review 2016: The Numbers and Trends*. Brussels: Centre for Research on the Epidemiology of Disasters (CRED).

Leading Edge Alliance (LEA). (2017) *Emerging Asian Economies*. Illinois: LEA.

The Oxford Dictionary. (2019) *Definition of Resilience in English*. Lexico. Available at: www.lexico.com/en/definition/resilience (accessed 10 June 2019).

UN-Habitat. (2019) *UN-Habitat: For a Better Urban Future*. Available at: https://unhabitat.org/resilience/ (accessed 10 June 2019).

United Nations. (2018) *World Population Prospects: The 2017 Revision*. New York: United Nations Population Division.

United Nations International Strategy for Disaster Reduction (UNISDR). (2015) *Global Assessment Report on Disaster Risk Reduction*. Geneva: UNISDR.

United Nations Office for Disaster Risk Reduction (UNDRR). (2019a) *Global Assessment Report on Disaster Risk Reduction*. Geneva: UNDRR.

United Nations Office for Disaster Risk Reduction/United Nations International Strategy for Disaster Reduction (UNDRR). (2019b) *Terminology on Disaster Risk Reduction*. Available at: www.unisdr.org/we/inform/terminology (accessed 15 July 2019).

United Nations, Economic and Social Commission for Asia and the Pacific (UNESCAP). (2015) *Asia-Pacific Disaster Report 2015: Disasters in the Asia and the Pacific: 2015 Year in Review*. UNESCAP. Available at: www.unescap.org/publication-series/asia-pacific-disaster-report

United Nations, Economic and Social Commission for Asia and the Pacific (UNESCAP). (2019) *Asia-Pacific Disaster Report 2019: The Disaster Riskscape Across Asia-Pacific, Pathways for Resilience, Inclusion and Empowerment*. Bangkok: UNESCAP.

Recurrent cyclones and floods

Institutional initiative for disaster resilience in Bangladesh

2 An old hand

Comprehensive disaster risk reduction institutional framework in Bangladesh

Introduction

This chapter explores disaster resilience issues in Bangladesh, one of the most disaster-prone and climate change-threatened countries in the world. Of the case study countries in this book, it is the most highly populated and largest in size (Briney, 2018; World Population Review, 2019a). Among these countries it is also the most highly impacted by disasters, in terms of human mortality, number of people affected and economic loss (Kafle, 2017). Thus, disaster resilience is a high priority of the government and a number of institutional initiatives have been undertaken in this regard, including the establishment of a dedicated ministry. This ministry dates back to the 1970s, hence the 'old hand' caption in this chapter's title. The United Nations Development Programme (2019) claims that 'With support from UNDP and others, Bangladesh became a leader in its institutional framework for disaster risk reduction and sustainable development, with a number of core government policies and programmes incorporating risk reduction from their earliest stages.' The case study in this chapter examines this claim by discussing one of the most significant disaster resilience programmes in Bangladesh.

Disasters continue to threaten the significant socio-economic progress achieved in Bangladesh in recent times. A number of hazards pose risk, among them recurrent floods and cyclones, although disaster resilience measures have significantly reduced human deaths, particularly by cyclones. Economic growth itself poses new challenges such as environmental problems, unplanned urbanisation and human-induced hazards. As evident from this chapter, the disaster risk scenario in Bangladesh is highly complex and continually transforming; furthermore, despite concerted effort, institutional policy and action face limitations.

The first part of the chapter outlines the key disaster resilience issues affecting the country, the prevalent and emerging hazards, administrative structure and institutional arrangements for addressing disaster risks. The second part is concerned with the case study of the Comprehensive Disaster Management Programme (CDMP), a significant programme that was implemented over more than a decade. Within this case study, a 'sub-case study' is discussed: a project implemented under CDMP that shows how the programme was translated into the ground. The case study programme and the project executed under it demonstrate both the achievements made in a developing country with serious constraints and the challenges inherent in such a context.

The research methodology for this chapter included two streams – a desktop literature review and empirical fieldwork. As there is quite a bit of literature relating to disasters in Bangladesh, hence the review was pointed towards identifying key aspects and exploring how the literature is aligned to them. The case study primarily refers to project documents and related publications. The field investigation consisted of interviews (and one focus group discussion at UNDP) of experts and others who had been associated with CDMP. The project under CDMP discussed here was explored at an earlier stage in a consultancy project, when it was possible to visit the project site and gather first-hand observations. A content analysis was performed on the literature and data on which this chapter is based.

Disaster context: recurrent cyclones and floods

Bangladesh is widely regarded as a highly disaster-prone country – two of its main hazards are tropical cyclones and riverine floods. Recurrent cyclones impacting the coastal areas can be ferocious and often lethal, and annual floods during the rainy monsoon season can have widespread, disruptive and damaging impacts throughout the country. The Comprehensive Disaster Management Programme (CDMP), the case study discussed in this chapter, was conceived as a response to major cyclone and flood events – mainly the 1991 and 1997 cyclones and the 1998 flood. CDMP II, the second phase of the programme, was informed by Cyclone Aila in 2009; an example of a project in this phase is also discussed within the CDMP case study.

Cyclones are typically triggered due to seasonal changes; in Bangladesh they tend to occur in the transitional period before or after the mid-year monsoons. The 700-kilometre-long coastline, together with offshore islands, is highly vulnerable to cyclones. More than 40 million people live in these coastal areas where poverty is extensive, further exacerbating vulnerability (Minar et al., 2013). The flat, low-lying coastal areas are often at sea level or minimally above it, with 62 per cent of the land below 3 metres in elevation (Center for Environmental and Geographic Information Services, 2009). The funnel shape of the Bay of Bengal and the flat deltaic topography are some of the main physical factors that contribute to the propensity for and impact of cyclones (Mallick and Rahman, 2013). The cyclones are often accompanied by high storm surges ranging from 1.5 to 9 metres; these can have devastating and long-term impacts, often exceeding the damage inflicted by the cyclonic wind (Dasgupta et al., 2011).

Bangladesh has a long history of frequently occurring cyclones. Among the cyclones in recent times, Cyclone Bhola in 1970 had the deadliest impact with about 500,000 lives lost. Since then, significant institutional risk reduction initiatives have followed, resulting in a 100-fold reduction in fatalities caused by cyclones (Haque et al., 2012). The 1991 cyclone was the second-most devastating cyclone – 138,000 people were killed, yet the significant reduction in deaths between these two cyclones is evident. Other major cyclones include Sidr (2007) which caused about 3,400 deaths (Government of Bangladesh, 2008) and Aila (2009) which caused 190 deaths (Khatun et al., 2017). More recently, in Cyclone Mora (2017) 18 people died (BBC News, 2017) and in Cyclone Fani (2019) five were killed (BBC News, 2019). These trends indicate the effectiveness of the different risk reduction measures implemented, including the widely acclaimed Cyclone Preparedness Programme (Bangladesh Red Crescent Society, 2019; United Nations Social and Economic Commission for the Asia Pacific, 2015) and cyclone shelters (Miyaji et al., 2017), among other factors such as increased awareness and an overall economic improvement leading to more sturdy buildings.

Cyclone Aila is notable because of the pronounced and prolonged impact of the storm surge beyond the cyclone itself. The cyclone struck the coast of south-western Bangladesh on 25 May 2009 – less than two years after the devastating Cyclone Sidr, which communities were still recovering from. More than 3 million people were affected in 11 coastal districts, with maximum impacts in Khulna and Satkhira districts. The turbulent cyclonic wind was accompanied by a storm surge higher than 6 metres, causing widespread damage. Along with 190 fatalities, thousands were injured and displaced and nearly 550,000 houses were destroyed or damaged. The extended coastal embankment of almost 2,000 kilometres was damaged, leading to inundation of vast areas of land – these remained under water for two to five years. Thousands of people became homeless and took shelter wherever they could. As well as the hardships of living in makeshift settlements, breakouts of epidemics such as diarrhoea and dysentery continued to exacerbate the suffering of the disaster-affected communities for a prolonged duration (for impacts of Aila, see BDRCS, 2011; Charlesworth and Ahmed, 2015; Hong Kong Red Cross, 2009; Islamic Relief, 2014; Khatun et al., 2017; Sadik et al., 2018). The example of the 'Disaster-Resilient Habitat' project discussed in this chapter was built after Cyclone Aila.

Bangladesh is known for its floods, recurring at regular intervals during the monsoons; there are numerous studies and media reports on floods in Bangladesh. Riverine floods and rainfall-induced floods, often in combination, are widespread; flash floods tend to be localised in areas with higher elevation such as in the eastern and south-eastern regions (Brammer, 1990; Kamal et al., 2018). The country's three main rivers – Padma (Ganges), Meghna and Jamuna (Brahmaputra) – and their numerous tributaries often spill over their banks, causing great damage and misery. Bangladesh is crisscrossed by numerous rivers that have created the largest riverine delta of the world (Misachi, 2017), the Ganges-Brahmaputra Delta. This is comprised primarily of flat, low-lying plainlands (Rashid, 1991); floods are an intrinsic part of the deltaic topography. Agriculture, especially rice cultivation, is reliant on the annual flooding, known as 'normal floods' and people have adapted to them over centuries. It is only when 'abnormal floods' occur beyond the regular flooding threshold that they result in a disaster (Rashid, 1993). It appears that such abnormal floods are increasing in frequency and magnitude because of climate change (Haustein et al., 2017; Parvin et al., 2016) and human action such as rapid and unplanned urbanisation and land use, upstream activities and construction of barrages (Lovelle, 2016). This contributes to an uncertain future scenario of floods in Bangladesh. Assessment of the physical magnitude and extent of floods is only one dimension of its impacts – its impact on people and their livelihoods in a context of deepening inequality (Byron and Parvez, 2019; Uttom and Rozario, 2019), as well as interaction with other hazards adds further dimensions to this complex phenomenon.

Data on the extent of flood inundation in Bangladesh over a 60-year period, from 1954 to 2014, shows that two of the major floods were in 1988 and 1998 – during these respective floods, 60 per cent and 68 per cent of the whole country went under water (Kamal et al., 2018). Massive floods also occurred in 1987, 2004 and 2007, as indicated by the data, and again subsequently in 2017 (George, 2017) and 2019 (Al Jazeera, 2019), thus indicating an increasing frequency of such heavy flood events. However, the 1998 flood was significantly greater in magnitude and impact; in addition to the widest extent of inundation, it lasted for 11 weeks, compared to the two weeks of the preceding largest flood of 1988 (Shah, 1999). The 1998 flood had extensive impacts, affecting about 30 million people, causing about 1,000 deaths and destroying over 500,000 houses, in addition to heavy damage to other key sectors such as agriculture and industries (Ahmed, 2000; Shah, 1999). As reported by officials working in government service at that time, this massive flood catalysed the Bangladesh government to explore risk reduction options as a shift from its previous approach that focused mainly on post-disaster response and relief provision. The inception of CDMP, the case study in this chapter, had its roots in this changing paradigm.

Country overview of key issues

Bangladesh is a riverine country composed of low-lying plains and river deltas, with a small proportion of hilly areas in the eastern and south-eastern regions. Its border is completely surrounded by India, except for a small south-eastern boundary with Myanmar, and is bounded by the Bay of Bengal to the south (see Figure 2.1). It is the largest country among the other South Asian countries in this book in terms of area and population.

The Rohingya refugee crisis – the mass migration of a Muslim minority group from Myanmar to nearby Southeast Asian countries – is a serious issue in Bangladesh. It is a complex problem that has become increasingly serious in recent times. Its dimensions and potential solutions are not entirely clear at this time (see for example The Guardian, 2019). As the disaster resilience needs of this community are of a specific nature it has not been covered here, with the view towards exploring it in future research. Other critical issues in Bangladesh that have strong linkages to disaster resilience in Bangladesh are discussed below.

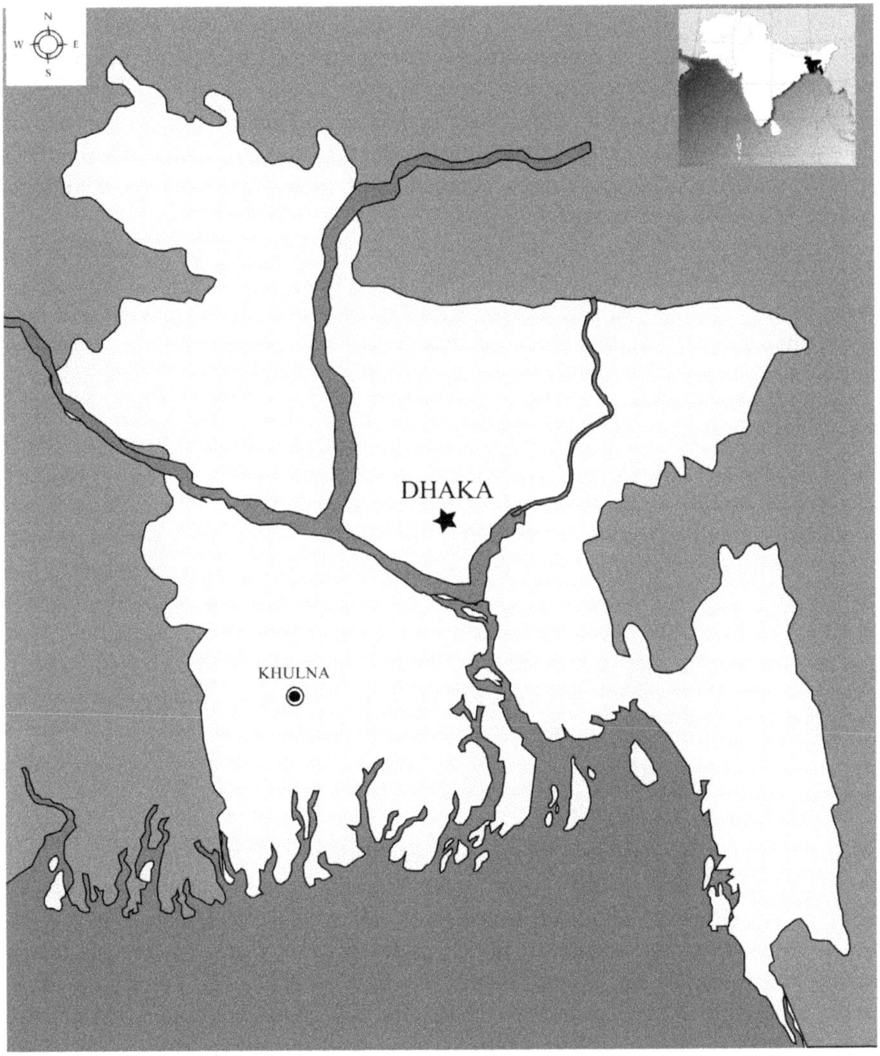

Figure 2.1 Map of Bangladesh showing locations relevant to this chapter

Population density

Bangladesh is a heavily populated country of more than 163 million people in a land area of 147,570 square kilometres, ranked eighth in the world in terms of population. It has the largest land area among the other countries discussed in this book – a little larger than Nepal but with almost six times the population (Briney, 2018; World Population Review, 2019a, 2019b). The average population density is close to 1,200 persons per square kilometre, but in some urban areas can be as high as 200,000 (Kabir et al., 2018). The largely flat topography with abundant rivers and wetlands provides ample opportunity for agriculture and fishing, hence settlements have been established widely throughout the country with an accompanying extensive population growth. Thus, there is intense pressure on and competition for resources – particularly land – impacting on the opportunities for economic development, particularly in the largely agrarian rural areas. This drives a continuous cycle of rural–urban migration.

In an attempt to diffuse the pressure on land, programmes were undertaken to settle landless populations in the relatively less populated Chittagong Hill Tracts in the south-east, but this resulted in serious conflict with and deprivation of indigenous hill dwellers (Halim and Chowdhury, 2015). There are also strong governmental and civil society family planning initiatives, resulting in some success by reducing the average fertility rate from 6.3 births per woman in the early 1970s to 2.3 in 2014 (National Institute of Population Research and Training et al., 2014). However, because of the large population base a significant growth of population in absolute numbers has continued. The large concentration of the Bangladeshi population entails significant vulnerability and exposure to hazards; despite recent institutional efforts to support disaster resilience, many people are frequently impacted by disasters and remain at risk. In addition to the human toll taken by disasters, the urban traffic congestion and heavy traffic throughout Bangladesh frequently results in accidents, often with tragic loss of many lives: in 2018, almost 7,800 deaths were caused by road, waterway, railway and aviation accidents (Bangladesh Passenger Welfare Organisation, 2019).

Urbanisation

As indicated by the statistics of the United Nations (2018), Bangladesh is still predominantly a rural country, with about 37 per cent of its population living in urban areas. However, rapid urbanisation is taking place and the urban population is expected to rise to almost 60 per cent by 2050. It has the highest urbanisation rate in South Asia – indeed in the whole of Asia. The great bulk of the urban population is concentrated in Dhaka, the capital, with a population of more than 18 million, accommodating almost one-third the urban population of the whole country and almost five times that of the next largest city, Chittagong (Trading Economics, 2019). It is among the top five fastest growing megacities in the world (Myers, 2016) and it is estimated that Dhaka's population density exceeds 40,000 persons per square kilometre, and up to 200,000 in some of the older southern parts of the city (Kabir et al., 2018) – among the highest population densities in the world (see Figure 2.2).

Bangladesh experiences an annual urban growth of almost 2 million persons (Worldometers, 2019) and it has been observed over the last decade or so that an estimated 300,000–400,000 mostly poor rural migrants continue to move to cities every year (Water and Sanitation for the Urban Poor, 2007). This movement is due to various 'push' factors including frequent disasters, increasing

Figure 2.2 Dhaka, the capital of Bangladesh, is one of the fastest growing cities in the world

pressure on land, rural unemployment and underemployment and rural–urban income disparity, as well as 'pull' factors such as employment opportunities and better living standards in the city (UN-Habitat, 2008; World Bank, 2007). The majority of new migrants in urban areas live in informal settlements or inadequate housing. The bulk of urban growth, however, is from the natural increase of the already large existing population, as well as conversion of agricultural land and wetlands in rural areas on the fringes (Shafi and Payne, 2007; UN-Habitat, 2009). Thus, Bangladesh is experiencing a state of constant transformation through a combination of demographic transitions and alteration of land use and the environment. Urban design, planning and delivery of services that both improve the quality of life for residents and make the expanding cities resilient to hazards should therefore be a priority.

Despite recent economic growth and overall reduction of poverty, income equality has widened in Bangladesh because of the concentration of wealth and inadequate distribution of resources (Byron and Parvez, 2019; Uttom and Rozario, 2019). The prevalence of poverty is widely evident; about a quarter of Bangladesh's population lives below the poverty line (Central Intelligence Agency, 2019) and about a quarter of Dhaka's population lives in informal settlements (Angeles et al., 2009; UN-Habitat, 2010). Urban areas characteristically include widespread informal settlements, large stocks of sub-standard buildings and extensive unplanned area developments. Poor drainage and waste management, traffic congestion, pollution and crime – among other such adversities – make urban life difficult to endure, particularly for women, children and the poor (for example, see Ahmed and Johnson, 2013; von Meding et al., 2018).

A high level of vulnerability results from this form of urbanisation. The growing and higher concentration of assets leads to greater economic losses. In a survey done around a decade ago, 78,000 buildings out of 326,000 in Dhaka were identified as vulnerable to disasters (Sarkar, 2009). The extent to which this has proliferated over the years in such a rapidly urbanising context can easily be imagined. In a study conducted by the authors and other researchers (Ahmed et al., 2018), it was found that there is a widespread disregard for building codes and many urban structures are built on soft land acquired by filling up water bodies; the majority of buildings are constructed by both a violation of planning regulations and avoidance of building codes. Many building collapses and fire incidents were noted in the study, the most prominent among them being the tragic 2013 Rana Plaza garment factory collapse. This structural failure accident killed more than 1,100 persons, the majority of whom were female workers. Natural hazards such as floods and storms can be seasonal, whereas such human-induced hazards persist on a day-to-day basis and constitute chronic vulnerability, exacerbating disasters when they occur. Such hazards have a more pronounced impact on the widespread urban poor communities who live in inadequate housing often in marginal areas. Poor water supply, sanitation and solid waste management, as well as substandard building materials contribute to their vulnerability and compound the impacts of hazards (Ahmed, 2014).

Environmental aspects

The nexus of high population density and a drive for economic growth, often with short-term profit motives, places great pressure on the environment; as a result, Bangladesh has a wide range of environmental problems. Only a few key ones can be discussed here.

The extensive construction of buildings as part of the rapid urbanisation process has led to a demand for building materials and land, both of which exert pressure on the environment. The construction sector has a great demand for sand, and despite regulations against it, sand is illegally extracted from rivers throughout Bangladesh, as is frequently reported in local media (for example, see Daily Star, 2019; Dhaka Tribune, 2018, 2019a). These reports point to powerful, politically connected persons conducting the sand extraction operations – local authorities are either powerless to stop it or are paid off. Such extraction creates flood risk by altering the river morphology, adding to the vulnerability of communities that are already exposed to frequent flooding.

Another issue related to the booming construction sector is the production of brick. For the vast majority of urban buildings, brick is the primary walling material; to cater to the demand in such a rapidly urbanising context, thousands

Figure 2.3 Brick kilns on the outskirts of Dhaka contribute to 60 per cent of the air pollution

of brick-burning kilns have been established around the major cities. This is a serious contributor to air pollution – it has been estimated by the Department of Environment that 60 per cent of the particulate pollution in and around Dhaka comes from brick kilns (Sachi and Price, 2018) (see Figure 2.3). Other environmental impacts include removal of topsoil to make bricks (Daily Star, 2018) and cutting down trees for fuel. Although coal is legislated as fuel, it is an imported product and hence expensive. There are hardly any alternatives to brick; although there have been some initiatives to popularise other materials such as concrete blocks (Dhaka Tribune, 2017), so far these have had limited application and are only available in the main cities. Other main materials used in urban construction are cement and steel – both are largely imported products with associated embodied energy impacts from transportation.

As Bangladesh is a riverine country, wetlands and water bodies of various types abound, nurturing great biodiversity and ecosystems. However, increasing demand for land has resulted in widespread filling of these water bodies: since 2010, 22 per cent of the existing wetlands around Dhaka have been lost (Hashim, 2019). There are a number of legislations and civil society initiatives to conserve wetlands (Haq, 2013), but these are widely violated (Hashim, 2019). There is also a range of other pressures on the wetlands such as discharge of industrial effluents, sewage and solid waste, overexploitation of aquatic resources and unplanned construction such as culverts and bridges – all of which have serious environmental repercussions (Bala, 2018; Haq, 2013). Importantly, wetlands serve as significant rainwater drainage areas – their loss results in floods and water logging, as found in many peri-urban areas that have been built by filling up wetlands (von Meding et al., 2018).

Increasing arsenic contamination of groundwater in Bangladesh is a major environmental problem. Out of 64 districts, 62 are seriously affected by arsenic contamination of drinking water from hand-pumps, known as 'tube-wells' (Ahmad et al., 2018). In some areas, 80 to 100 per cent of the tube-wells are contaminated with arsenic; even 20 years after the problem was detected, there are 20 million rural people drinking this contaminated water (Human Rights Watch, 2016) exposing them to Arsenicosis, an illness that leads to a range of health complications and also fatality (Ahmad et al., 2018). Despite various institutional mitigation initiatives to provide safe drinking water, it appears that affordable and sustainable solutions at the scale that is required remain a challenge (Islam and Islam, 2019).

Disaster risk profile

Bangladesh is highly disaster-prone and throughout its existence has been shaped by the impact of hazards, both extensive (low-severity, high-frequency events) and intensive (high-severity, mid-frequency to low-frequency events). As a result, managing disasters and their impact has been a major area of focus, with investments in disaster risk reduction and management significantly decreasing disaster mortality, as mentioned above for the case of cyclones in the disaster context section. Nonetheless, economic losses resulting from disasters continue to increase, presenting a challenge to national development (International Federation of Red Cross and Red Crescent Societies, 2018). With the different changes happening globally and in Bangladesh, a transforming disaster profile is emerging: Bangladesh is one of the world's most at-risk countries facing the advancing phenomenon of climate change (Kreft et al., 2014). There is also the risk of earthquakes, which pose a challenge particularly for Bangladesh's rapidly growing cities (Islam et al., 2016).

These emerging risks present major challenges to the continued human development, poverty reduction and economic growth of the country, and to the lives, livelihoods and health of its people. The poorest, most marginalised and vulnerable communities are often hardest hit by disasters in Bangladesh as they are repeatedly exposed to hazards without the means to recover adequately. Disasters are a major threat to people's life and livelihoods in Bangladesh; in a very short period of time, a disaster can eradicate years of development effort.

Despite these challenges, Bangladesh has made major gains in improving socio-economic conditions in recent years with positive economic trends, accelerating growth and improved indicators of human development. The country has made significant achievements in the Millennium Development Goals

(Bangladesh Planning Commission, 2012) and is making good progress with the Sustainable Development Goals (BPC, 2018). With an average 6 per cent economic growth in the last decade, it reached lower middle income country status in 2015 (World Bank, 2019). As an ascending middle income country, Bangladesh has entered a new development context, with a growing asset base and connectedness to global markets. However, the national economy is at risk from disaster impacts and climate stresses and a large amount of gross domestic product (GDP) is lost each year due to these events. In the decade from 2005 to 2014, economic losses due to disasters amounted to more than USD285 million – between 0.8 to 1.1 per cent of GDP (Emergency Events Database, 2015; PreventionWeb, 2014) and it is among the top 15 most vulnerable countries in the world (UNESCAP, 2015). The industrial sectors are highly vulnerable to economic losses: the garment sector, vital to the national economy, incurs major losses in floods (Sebastio, 2018) and is particularly at risk from earthquakes because most of the garment factories are located in multi-storey buildings.

The key hazards in Bangladesh – cyclones and floods – have been discussed above in the section on the disaster context because of their direct bearing on the inception of the case study programme in this chapter. Some of the other hazards are discussed below.

Earthquakes

Earthquakes pose one of the highest risks in Bangladesh because of the country's location in a tectonically active region. Some of the major cities including Dhaka, Chittagong and Sylhet are at risk of massive destruction by earthquakes from nearby seismic faults to the east (Miller, 2016). Dhaka is located in a moderate earthquake risk zone, but experts have been warning of severe consequences if a major earthquake does strike this densely built and populated city (Hindustan Times, 2010; New Age, 2011). Because of unplanned urbanisation and sub-standard building practices where building codes and micro zonation maps are not followed, one prominent earthquake expert has suggested that 'Dhaka is among the most vulnerable megacities in the world' (J.R. Choudhury in the New Age, 2011). Although Bangladesh has not experienced a major earthquake in recent times, strong tremors often occur, creating public fear (Daily Star, 2016; Iaminweb, 2010; U.S. Geological Survey, 2012); a massive earthquake is a real and significant possibility. In addition to causing a devastating death toll, a major earthquake would also lead to huge economic losses. With urban industrial production facilities and road networks likely to be damaged or blocked by debris, lengthy disruption of business can occur. It can be expected that, with the urban private sector being the major driver of national economic growth, protracted disruption of production due to the shock of a major earthquake would have a significant, long-lasting negative impact on national economic growth and poverty reduction.

One of the main dangers arising from earthquakes is the collapse of buildings, which can kill or trap people, and which is more acute in the case of multi-storey buildings – particularly those built of heavy materials such as brick and concrete, as is widely prevalent in Bangladesh's urban and peri-urban areas. Recognising this, the Bangladesh National Building Code (BNBC) was updated in 2015 (Housing and Building Research Institute, 2015). Unlike urban areas, most houses in rural areas and in informal settlements are built of light-weight materials such as metal sheet, timber and bamboo, and are usually only one or two storeys high. Thus, the risk of getting killed or seriously injured in an earthquake from such buildings is not very high. However, brick is increasingly being used to build walls of rural and informal houses, which are often poorly constructed and lack a reinforcing structure. Additionally, old dilapidated buildings are often occupied. Such buildings pose a risk in the event that a strong earthquake strikes. The densely built informal settlements with typically narrow roads amplify the risk because of the difficulty they pose for rescue operations.

Riverbank erosion

Riverbank erosion is a common and serious problem in Bangladesh due to the deltaic topography (see Figure 2.4) and it has conceivably been increasing with

Figure 2.4 An area affected by riverbank erosion

climate change (Displacement Solutions, 2012; Start Fund Bangladesh, 2019). Riverbank erosion has been forcing people to migrate or resettle and has rendered millions homeless. The majority of slum dwellers in large urban and metropolitan towns and cities are victims of erosion (Islam and Rashid, 2012; Rahman, 2010). The major rivers like the Jamuna, Ganges, Padma, and the Lower Meghna, Arial Khan and Teesta rivers are highly erosion-prone (Green Bangla, 2013). Structural interventions are costly and need to be complemented by non-structural measures, such as erosion prediction, warning and evacuation. Riverbank erosion is an institutional priority area and there are several initiatives to manage it and address its impacts (for example, see Asian Development Bank, 2013; Bangladesh Water Development Board, 2014).

Drought

As discussed above, Bangladesh is known for its floods, but less widely discussed is the fact that large parts of the country also suffer from drought annually in the dry season, particularly the north and north-western regions (Alamgir et al., 2015; Dey et al., 2011). A combination of climate change and variability, along with human factors such as exploitation of water resources and deforestation contribute to a pattern of increasing severity of drought episodes (Center for Environmental and Geographic Information Services, 2013; Rahman and Lateh, 2016). The impacts on agriculture are often critical – particularly on rice cultivation, the most important crop in Bangladesh. This has a range of impacts, especially on human health and well-being in terms of nutrition (Dey et al., 2011). Famine-like conditions, locally known as 'monga', often occur in the drought-prone areas during the lean agricultural seasons (Paul et al., 2013). The drought-prone northern areas are also affected by cold snaps with severe impacts on the population, in particular on the poor (Hasnat, 2018). Thus, Bangladesh experiences a dual-edged impact of disasters: excessive water causing floods in the wet season and lack of water causing drought in the dry season. This means that institutional attention to disasters is required throughout the year.

Windstorms/tornadoes

While cyclones affect the coastal areas, inland seasonal north-westerly windstorms are common in Bangladesh, usually during the pre-monsoon period of March to May, some of which become tornadoes. There are also cases of windstorms occurring in other times of the year, which do not follow the typical seasonal pattern – possibly due to climate change (Finch and Dewan, 2007). A study indicated that in 150 years since 1865, 116 tornadoes struck Bangladesh causing more than 9,500 deaths (Hosen and Jubayer, 2016). According to the World Meteorological Organization, one of the deadliest tornadoes in the world hit Bangladesh in 1989, killing 1,300 people and injuring more than 12,000 (Dolce, 2014). Often the windstorms/tornadoes are accompanied by severe hail and lightning which can cause great damage and harm to people

(BDNews24, 2014). For example, two persons were killed by lightning in a tornado in 2018 (Islam, 2018); this is a particular risk in rural areas where people work outdoors. In urban informal settlements and rural areas, even minor windstorms can cause havoc because of the typically weak house construction. The commonly used corrugated iron roofing sheets get blown off, wind-driven rain penetrates inside and damages household assets, and the residents' long-term well-being is affected.

Landslides

Bangladesh is characterised by its predominantly flat topography, but hilly areas comprise about 12 per cent of the country (USAID, 2010), particularly in the south-eastern Chittagong Hill Districts (CHD). In these areas, landslides have emerged as a major hazard, usually triggered by heavy rainfall. Rain-induced flash flooding is often accompanied by landslides in the hilly terrain, presenting a complex multi-hazard scenario. Because Chittagong is a seismically active zone (Miller, 2016), there is also the potential for earthquakes to trigger landslides. Human settlements – many of which house low-income communities – are built on hill slopes by cutting into the hills (Ahmed, 2007, 2017). Depletion of topsoil vegetation and denuding the hills for building settlements contributes towards soil loosening and erosion of the hills. After heavy rain large chunks of hilly soil become dislodged and slide down, causing devastation in the dense settlements on the slopes and at the bottom. In some of the cases, the surrounding areas become flooded and escape routes are blocked, resulting in people becoming trapped and getting buried by landslides. In some cases, people – especially children – are washed away in the turbid flood currents (Ahmed, 2007). Such a multi-hazard situation presents tremendous challenges for institutions and at-risk communities in Chittagong, many of which continue to live at risk in hilly locations. This is a hazard that is drawing institutional attention, and there are some initiatives for early warning, as implemented in the CDMP case study project in this chapter.

Tsunamis

Bangladesh was not seriously impacted by the 2004 Indian Ocean Tsunami, beyond two persons who were killed while out at sea; the only previous recorded tsunami in Bangladesh was in 1762 (WorldData, n.d.). Hence it is categorised as a medium-level hazard (Global Facility for Disaster Risk Reduction, n.d.). However, the 2004 tsunami raised awareness of the hazard, and the Bangladesh Meteorological Department monitors tsunami risk and early warning. With climate change induced sea level rise, the areas of tsunami risk are expected to increase (GFDRR, n.d.), requiring coastal planning to take this hazard into account.

Human-induced hazards

Industrial hazards occur regularly in Bangladesh. Many chemical factories and warehouses are located in residential areas, making densely built urban areas highly vulnerable. Chemical explosion is another incident common in industrial premises. On average 21 people die in Bangladesh due to toxic chemicals each month (Rozario and Uttom, 2014). Spill of oils and toxic chemicals from maritime vessels is also on the rise – the 70,000-litre oil spill near the Sundarbans protected forest area in 2014 caused serious damage to the flora and fauna and environment (Sunny, 2017). Sub-standard construction and non-compliance with building regulations cause frequent collapse of buildings in Bangladesh (Ahmed et al., 2018) (see the sub-section on urbanisation above). Major fires occur frequently, causing huge loss of lives and properties. Nearly 2,000 people were killed in about 200,000 fire incidents between 2004 and 2018 and fire incidents have tripled over the last two decades (Dhaka Tribune, 2019b). Fires are often a result of non-compliance to building codes (Ahmed et al., 2018). The Bangladesh Fire Service and Civil Defense (BFSCD) is trained and equipped, but often hindered by traffic congestion in the provision of timely response to incidents.

Disasters are often the outcome of inadequate development choices or gross capacity gaps to implement sound policies and programmes to reduce risk. Catastrophic disasters resulting from human-induced hazards in Bangladesh have shown how disaster risk and poverty are closely interlinked.

Climate change

Bangladesh is one of the world's countries most vulnerable to climate change (Inform, 2019; Institute of Economics and Peace, 2019). Some of the negative impacts of climate change include increases in incidence and intensity of extreme weather events and hazards such as rising sea levels and soil salinisation. With its low elevation, Bangladesh is vulnerable to widespread inundation by a small rise in sea level induced by climate change (Glennon, 2017). Salinity intrusion has already become a serious problem in coastal areas (World Bank, 2015), posing a threat to ecosystems, livelihoods and public health and diminishing access to freshwater for household and commercial use. More than a decade ago, about 20 million people in the coastal areas of Bangladesh were already affected by salinity in their drinking water (Ministry of Environment and Forest, 2006). Bangladesh Water Development Board (BWDB) undertook studies on groundwater availability and found that in some coastal districts there was no freshwater layer as deep as 300 metres; in many cases, saline water was found in aquifers at 200 metres (Zahid, 2017).

Another key impact of climate change is heat stress. An available 10-year record indicates a consistent rise of average temperature (Rabbani, 2010). In addition, increasing urban development and traffic continue to contribute to the 'urban heat island effect' (Das and Karmakar, 2015) by generating more heat and adding to the climate change-induced temperature rise. Once again, this has a range of implications including loss of productivity and income, negative impacts on health and disrupted water and energy supplies. Such a situation is conducive for the ignition and rapid spread of urban fires, as discussed above in the sub-section on human-induced hazards.

As Bangladesh is a low-lying country with numerous rivers and wetlands and has a tropical monsoonal climate, the country has historically experienced natural floods, as discussed above in the section on the disaster context. However, climate change has resulted in erratic weather patterns and increased frequency and intensity of extreme weather events – as evident globally – with the greatest impacts on developing countries, particularly on the poor (Intergovernmental Panel on Climate Change, 2012). The main cities, especially Dhaka, continue to draw climate change refugees from rural areas (Displacement Solutions, 2012), fuelling its population growth and adding to its existing widespread poverty. Despite the government's various initiatives to address climate change, the urban poor are generally neglected because of their uncertain tenure status in informal settlements (Banks et al., 2011). The impact of floods and water logging and associated secondary effects on the urban poor who live in settlements in marginal areas without drainage has been mentioned above in the sub-section on urbanisation; these impacts can be expected to magnify in the future. A study on the effects of climate change on the urban poor in Dhaka indicated serious health impacts, such as fever, and disruption of water services (Khan, 2010).

Climate change poses risks to national development, requiring climate change adaptation to be integrated with relevant disaster resilience policies and strategies as well as in the socio-economic development process, natural resources management and livelihoods support efforts. Since poverty, sustainable development, disasters and climate change are closely interlinked, it is crucial to integrate resilience measures in development initiatives where people's livelihoods are at risk.

Policy aspects of disaster resilience

Bangladesh has historically been frequently affected by disasters, particularly recurrent cyclones and floods, hence there is strong institutional commitment to achieve disaster resilience. The country is somewhat exceptional in having a dedicated ministry to deal with disasters, and other than Sri Lanka, none of the other South Asian countries in this book have such a ministry. Compared to the Sri Lankan ministry that was established in 2005 after the Indian Ocean Tsunami, the Bangladeshi ministry was established long before – in 1972, soon after Bangladesh became an independent country in 1971. In that sense, Bangladesh is an 'old hand' in the disaster resilience field.

Because of this longstanding institutional engagement there is an extensive and complex network of institutions working in this domain, starting from the top national level down to the lowest tier of local administration. In addition, disaster resilience is mainstreamed such that many other line ministries have sector-specific programmes dealing with disasters. Furthermore, Bangladesh has a large network of national and international non-governmental organisations (NGOs), and together with UN agencies, they are also widely engaged in this field. Documentation of this complex network is provided in a reference handbook by the Center for Excellence in Disaster Management & Humanitarian Assistance (2017). In addition to information obtained from fieldwork, several sources have been consulted to provide the overview below on the institutional structure and policy framework for disaster resilience (Ahmed, 2019; Alam and Ray-Bennett, 2019; Bangladesh National Parliament, 2012; CEDMHA, 2017; Islam, 2013; Rezoane, 2016; websites of the Ministry of Disaster Management and Relief and Department of Disaster Management).

Administrative structure

The national government bodies are based in the capital, Dhaka, with a range of local level administrative bodies throughout the country (for a description of the Bangladesh administrative structure, see Rokon, 2013; Wikipedia, 2019). There are eight geographical divisions, which are further sub-divided into 64 districts (*zila*). Under the districts, there are 492 sub-districts (*upazila*). From this point, the administrative structure becomes divided according to urban and rural jurisdictions. In rural areas, unions are the administrative units, which are governed by Union Councils. In urban areas, the smaller towns are known as Municipalities (*pourashabha*) and the larger metropolitan cities have City Corporations; the large size of Dhaka has led to forming two corporations – North and South. The unions and city corporations are comprised of the final smallest administrative unit, wards. Such a complex network of a hierarchy of administrative units is necessary for the governance of a highly populous country such as Bangladesh.

Institutional arrangement for disaster resilience

The institutional arrangement for disaster resilience in Bangladesh has an elaborate structure, consisting of a large number of concerned institutions, as indicated by CEDMHA (2017) and Haque and Uddin (2013). For the sake of this chapter and to relate it to the administrative structure discussed above, a somewhat simplified organogram adapted from Islam (2013) is shown in Figure 2.5. There is a related regulatory framework that defines the relationship between the different policy instruments in tandem with the institutional structure (see for example, Alam and Ray-Bennett, 2019; Islam, 2013; Rezoane, 2016).

The National Disaster Management Council (NDMC), headed by the Prime Minister, is the supreme body for providing overall direction for disaster resilience which includes disaster risk reduction, mitigation, preparedness, response and recovery. NDMC is guided by the National Disaster Management Advisory Council (NDMAC). Because of the cross-cutting nature of disasters that have a bearing on almost all the sectors, there is an Interministerial Disaster Management Coordination Committee (IMDMCC). This is chaired by the Ministry of

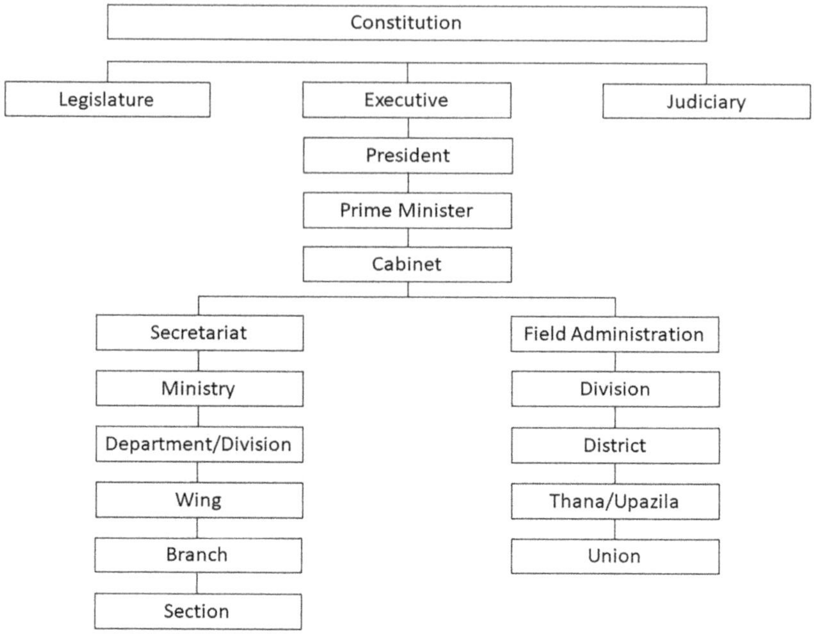

Figure 2.5 Institutional structure for disaster resilience in Bangladesh (adapted from Islam, 2013)

Disaster Management and Relief (MoDMR), and composed of members drawn from all the key ministries, as well as the Bangladesh Red Crescent Society which manages the vital Cyclone Preparedness Programme (see for example Ahmed, 2007). The IMDMCC plays an important role by reviewing the national relevance of the work conducted under MoDMR. MoDMR has the prime responsibility for translating the government's directives and coordinating national disaster resilience efforts across all agencies. As disaster resilience is a multi-sectoral and multifunctional field, functional and hazard-specific planning and execution responsibilities are vested in agencies with primary technical/management focus related to specific sectors, whereas MoDMR has an overall coordinating and facilitating role as 'Secretariat' to NDMC.

With support from the Inter-Ministerial Disaster Management Coordination Committee (IMDMCC) and NDMAC, NDMC coordinates disaster-related activities at the national level. Coordination at district, sub-district and union levels is done by the respective local level Disaster Management Committees (DMCs) under the Department of Disaster Management of MoDMR. A series of inter-related institutions at both national and sub-national levels operate to implement planning and coordination of disaster risk reduction and emergency response management.

Policy instruments and challenges

There is a wide range of policy instruments in Bangladesh relating to disaster resilience, including those dealing with the related field of climate change. Here, only those instruments concerned directly with disasters and belonging to the portfolio of MoDMR are discussed. The policy instruments for disaster resilience are related to a regulative framework consisting of policy and legislative aspects, and best practice guidelines (Alam and Ray-Bennett, 2019; Islam, 2013; Rezoane, 2016). Within that framework, the main policy instruments include:

- *Standing Orders on Disaster* (SOD), issued by the Ministry of Food and Disaster Management (which was changed to MoDMR in 2012) in 1997. This was an important milestone towards guiding and monitoring disaster-related activities in Bangladesh. It was updated in 2010. The SOD includes details of the specific roles and responsibilities of the different institutions dealing with disasters.
- *Disaster Management Act 2012*, which has the objectives of a substantial reduction of disaster risk to an acceptable level with appropriate interventions. This act of the Government of Bangladesh endorses the SOD and provides the legal basis for undertaking actions for disaster resilience in the country.
- *Disaster Management Policy 2015*, a strategic policy instrument in line with wider national objectives. It deals with both disaster risk reduction and emergency management, and provides the national perspective and principles to define disaster resilience related efforts.
- *National Plans for Disaster Management* (NPDM), a series of planning instruments for five-year periods. NPDM 2010–2015 was the first policy document of its kind and reflected a paradigm shift from disaster relief and response to a comprehensive risk reduction culture, with emphasis on capacity strengthening. It helped achieve several milestones leading to the Disaster Management Act (2012). It had an approach based on hazard-specific and sectoral plans, translated into local level plans at the different administrative tiers. The subsequent NPDM 2016–2020 provides policy direction in terms of national level action plans in alignment with the global *Sendai Framework for Disaster Risk Reduction* (SFDRR). These action plans have indicative time frames over the subsequent five years and 34 core targets, to be continued until 2030.

In terms of institutional arrangements and policy instruments, Bangladesh has a well-established and elaborate framework, which has been developed over a long time, evident from a dedicated ministry for this purpose. Yet, during the fieldwork, several disaster resilience experts revealed a number of shortcomings in the public sector. They pointed out that there are numerous projects implemented in Bangladesh, especially by NGOs, but the government fails to utilise the lessons from these projects and mainstream and upscale them in its programmes and actions. A knowledge-based approach seems to be lacking; even when new knowledge is generated, it is not used to inform policy and practice. Many new ideas and initiatives are generated, evident from the many policy instruments developed, but they appear not to be sustained and built upon to take to another level by mainstreaming it. As observed by an expert, 'The government doesn't have the mechanism to integrate such knowledge into its work.'

A significant amount of funds is spent on disaster resilience, but the experts suggested that there is a need for a thorough investigation on the extent these efforts are mainstreamed and scaled up. There is excessive reliance on funding from international donors; many of these donors encourage the government's ownership of the projects implemented by their funding, but the government lacks the capacity to make progress beyond the projects' time frame. This was captured by a succinct comment: 'There is no long-term view, they behave more like service providers on a short-term project.' It seems that the interest is on capital-intensive projects such as cyclone-shelters, embankments and culverts that allow opportunity for siphoning off funds. An example was provided of the construction of 20-metre high culverts to drain off storm surges, but there was very little study on whether such a height was required in that particular context.

Thus, many of the policy instruments remain simply as documents without much application. It seems that government officials brag in international forums about their extensive policy instruments, but have little to show in terms of translating them on the ground. The NPDM 2016–2020 is a case in point. More than two years after it was produced, and well into its planned timeline, it was only available online as a draft version on MoDMR's website. It seems that government officials consider NPDM 2016–2020 as a 'live document' by virtue of the draft being available on the Internet, and perhaps this is the eventual extent of its promulgation. Thus, while the planning process behind production of the document and the product itself were significantly rigorous, its promulgation in terms of translating into implementation, institutional capacity building and uptake at different levels remained uncertain. The process leading up to the production of the plan was open and transparent, with active involvement of a wide range of stakeholders. However, after it was produced the process became largely opaque and closed and engagement with the public domain became limited. None of its specified target actions were implemented (Ahmed, 2019).

This is a key challenge in the institutional arena in Bangladesh: to extend beyond production of policy documents, to actually implement disaster resilience initiatives informed by the documents. The case study of CDMP in this chapter also evinces this challenge.

Case study: Comprehensive Disaster Management Programme (CDMP)

The Comprehensive Disaster Management Programme (CDMP) is a unique initiative of MoDMR (previously MoFDM) and one of the largest such programmes globally. It has included a wide range of projects and activities throughout all the disaster-prone areas of Bangladesh for more than a decade. It was implemented in two phases: Phase I (2004–2009) and Phase II (2010–2015) and consisted of a strategic institutional approach (Islam, 2013) of working with more than a hundred partners including government agencies, NGOs, UN agencies, academia and a wide range of other stakeholders. Most importantly, it aimed to achieve a 'paradigm shift' (MoDMR, 2015, p. 8; UNDP, n.d.(b), p. 1) from a predominantly reactive post-disaster response and rehabilitation-orientated approach to disaster risk reduction and building resilience (Halder, 2010). CDMP is characterised by its holistic and integrative orientation consisting of a 'multi-hazard, multi-sector and multi-stakeholder' (Halder, 2010; Policy Programme & Partnership Development Unit, 2009) programme of action.

CDMP is widely acclaimed as one of the first initiatives to strongly advocate for a focus on disaster risk reduction beyond only emergency management. Because of the wide range of stakeholders involved in the programme, this message has spread throughout the disaster resilience community of practice in Bangladesh. The programme made significant contribution to national disaster resilience policy: the drafts of the Disaster Management Act and Disaster Management Policy were updated and finalised; the Standing Orders on Disasters, originally produced in 1997, were updated from largely a response focus to incorporating a risk reduction approach; and the National Plan for Disaster Management 2010–2015 was developed. This was achieved through the backing provided by CDMP to the government (Halder, 2010). Even the name and functions of the previous Ministry of Food and Disaster Management (MoFDM) was changed to the Ministry of Disaster Management and Relief (MoDMR) to maintain an exclusive focus on disasters. The valuable contribution of CDMP to the disaster resilience field during its term is widely recognised by experts and professionals in different agencies in Bangladesh.

Background to the programme

Massive cyclones in 1991 and 1997 presented immense challenges to the government, despite having reduced the larger extent of risk and impacts of the past. Also, as discussed above in the section on the disaster context, the flood of 1998 – one of the heaviest in previous history – catalysed the government to initiate action and seek funding to reduce the risk of such disasters. As narrated by personnel working in CDMP I at its early stage, the programme had its roots in these disasters and the consequent intent of the government to reduce long-term disaster risk across the country. While CDMP I was ongoing, two strong cyclones – Sidr (2007) and Aila (2009) – impacted Bangladesh, with the latter causing extensive long-term damage. These disasters, particularly Cyclone Aila, contributed towards the formulation of its second phase CDMP II (Policy, Programme & Partnership Development Unit, 2009).

Key features of the programme

In the first phase, CDMP I (2004–2009), national systems and policies relating to disaster resilience were reviewed and updated. The focus was on policy level work and building capacity within the government. It established a model for disaster risk reduction through 'defining and redefining the risk environment' and thereby 'managing the risk environment'; and integrating it with emergency management by 'responding to the threat environment' (Ministry of Food and Disaster Management, n.d., p. 9). This is reflected in the five strategic focus areas of CDMP I as listed by MoFDM (n.d.):

'i. Professionalising the disaster management system;
ii. Mainstreaming disaster risk management within development and investment planning processes;
iii. Strengthening community institutional support systems;
iv. Expanding mitigation and preparedness programmes to cover a wider range of hazards and geographical areas;
v. Operationalising response management systems.'

Thus, while bringing to the forefront the agenda of advancing a disaster risk reduction paradigm shift, the necessity and importance of emergency management was also recognised in the programme and provisions made for it. In that sense, it was a comprehensive initiative as the programme's title indicates.

The above strategic focus areas are reflected in five sub-programmes, each with its components of action: Capacity building; Partnership development; Community empowerment; Research information management; and Response management. Each is funded by three main international donor agencies, in combination or individually, and also included some funding from the Government of Bangladesh: UK Department for International Development (DFID), European Union (EU) and United Nations Development Programme (UNDP) (Halder, 2010). The funding was channelled through UNDP, which managed the funds. The total budget was US$27.12 million (Russell et al., 2009).

CDMP I originally targeted seven severely disaster-prone districts; after Cyclone Sidr in 2007 it was expanded to nine more districts to include cyclone-affected coastal areas, and also some flood-affected inland areas (Halder, 2010). Because of its comprehensive nature, a wide range of activities were undertaken. It is beyond the scope to list them here; only a few key elements can be mentioned, as below:

- Community Risk Assessment (CRA), an important methodology established in CDMP I, which continued to be applied in CDMP II widely throughout Bangladesh. Many NGOs have adopted the CRA tool and utilise it independently. The CRA is applied to develop Risk Reduction Action Plans (RRAPs) at the local level.
- The Local Disaster Risk Reduction Fund (LDRRF), a mechanism established to support projects at the local level based on RRAPs, often targeted

for the most vulnerable and poorest communities (UNDP, n.d.(b)). The LDRRF was expanded in CDMP II, but even in CDMP I it was estimated to have benefitted over half a million people through more than 500 projects (Russell et al., 2009).
- A national Disaster Management Information Centre (DMIC), established with 24/7 operational capacity to monitor and report hazards that pose risk, allowing for the dissemination of early warning throughout Bangladesh.
- A Climate Change Cell, established under the Ministry of Environment and Forests, with a view towards integrating disaster risk reduction with climate change adaptation and linking related national policies.

CDMP I was considered successful by the government (MoDMR, 2015). During its course, eight evaluations, including a terminal evaluation and four audits, were conducted – all of which generally indicated positive results (Russell et al., 2009). Hence the second phase, CDMP II, was initiated and received substantial support from donor agencies and the government. In fact, the programme was designed to comprise two phases, with CDMP I as a pilot phase, serving as a foundation to expand into a much larger phase in terms of funding and activities.

CDMP II was originally planned for 2010 to 2014, and was later given a no-cost 1-year extension. Implemented by MoDMR, it was funded by a number of donors including the Government of Bangladesh, UK Department for International Development (DFID), European Union (EU), Norwegian Agency for Development Cooperation (NORAD), Swedish International Development Cooperation Agency (SIDA), Australian Agency for International Development (ex AusAID) and UNDP. The total budget was US$76.32 million (United Nations Development Programme (UNDP), n.d.(a)), nearly three times more than the budget of CDMP I, indicating the expanded scale of this phase. The programme was implemented in 40 vulnerable districts – much more widespread than the 16 districts covered in CDMP I. It focused on building the capacity of the Disaster Management Committees at different administrative levels down to the lowest union level.

As discussed in the CDMP II Project Completion Report (MoDMR, 2015), there were seven main outcome areas:

- 'Professionalising disaster management': Some of the key activities included establishing educational programmes relating to disaster resilience, development of knowledge products available through an e-library and sensitising journalists for more effective reporting on disasters.
- 'Rural risk reduction': Bangladesh still being a largely rural nation, significant investments were made for structural and non-structural disaster risk reduction measures in rural areas (see Figure 2.6). Almost half of the CDMP II budget was invested in LDRRF projects in rural areas benefitting more than 3 million people.

Figure 2.6 Repairing embankments after Cyclone Aila through a cash-for-work programme under CDMP II (courtesy of Sanjib Kumar Saha)

- 'Urban risk reduction': The rapidly urbanising context of Bangladesh presents a host of emerging risks – these were addressed through a combination of structural and non-structural measures. While in CDMP I the focus was on earthquakes in three cities, CDMP II followed an all-hazards approach to prepare risk assessments for six cities.
- 'Preparedness and response': More than 30,000 volunteers were trained in cyclone and flood preparedness programmes, and more than 5,000 secondary school teachers were trained in earthquake preparedness. Early warning systems were significantly expanded.
- 'Mainstreaming disaster risk reduction and climate change adaptation': Capacity was built on an integrated approach to disaster risk reduction and climate change adaptation at different levels from government staff to farmers. Importantly, the knowledge was incorporated into school curricula, including textbooks from year 3 to 12.
- 'Managing community level adaptation to disaster risk from a changing climate': With a focus on capacity building, governmental and other professionals were trained and provided training manuals and adaptation toolkits. A number of research projects were supported to study climate change risks and impacts.

There was a wide range of other activities including providing training and equipment to the Fire Service and Civil Defence (FSCD) agency and support for Information and Communication Technologies (ICT) for disaster resilience, including expanding the DMIC to the sub-district level. CDMP II represents a large investment for disaster resilience in Bangladesh and virtually touched almost all related institutions, organisations and communities in Bangladesh.

An example of a CDMP II project: 'disaster resilient habitat'

While most of the projects undertaken through CDMP focused on policy and capacity building, its housing and settlement development projects – termed 'disaster resilient' (MoDMR, n.d., p. 1; Nahar et al., 2015, p. 3; United Nations Development Programme (UNDP), n.d.(b), p. 3) – provide evidence of visible and tangible outputs that are readily conducive for field investigations. An example of such a project is discussed below.

Background to the project

From the 1960s, a system of embankments was built along the southern Bangladeshi coast to provide protection from storm surges (Saari and Rahman, 2003). However, since then the areas outside the embankments were widely settled by landless and marginal communities, exposing them to cyclone impacts. Such settlements are in an extremely precarious situation – uncertain tenure and low incomes disallow the construction of durable dwellings, thus a disaster can cause severe devastation because of the exposure and the weak nature of the housing. In 2009 Cyclone Aila caused widespread devastation to informal settlements such as these. Additionally, breaches of the embankments by the storm surge resulted in vast areas inside the embankments becoming inundated by saline water. For more than two years after this event, people built makeshift shelters on embankments or any available high land and lived there under very difficult conditions due to delays in repairing the embankments and draining the water. Huge loss of housing, agriculture and other assets was experienced, encapsulated in a comment by a local resident, 'We turned destitute within the blink of an eye' (Nahar et al., 2015, p. 10).

The coastal areas of the southern district of Khulna were severely impacted by Cyclone Aila. In the village of Bainpara, which is located in a remote area and comprised of 208 households, many people took shelter on embankments or migrated elsewhere after Cyclone Aila. In 2011, under CDMP II, a project to rehabilitate the community was initiated. The local Union Disaster Management Committee (UDMC) identified the 84 poorest and most vulnerable households, out of which 26 households had already left the area. The remaining community of 58 households were living together in a makeshift settlement and were selected as beneficiaries of the project for establishing a new settlement. Land was procured in a nearby location and the community was resettled there with housing and related facilities. This 'Disaster Resilient Habitat' project was completed in 2014. Details of the project have been gathered from MoDMR (n. d.) and Nahar et al. (2015); the project was also looked at under a larger study by one of the authors (see Ahmed, 2016).

Main project partners

A number of partners were involved in the project: as mentioned above, the local UDMC carried out the beneficiary selection and was coordinated by the sub-district (*upazila*) Disaster Management Committee (UzDMC); the governmental Housing and Building Research Institute (HBRI) provided the prototype house design and construction specifications through consultations with the Bangladesh University of Engineering and Technology (BUET); the NGO, Centre for Natural Resources Studies (CNRS), undertook the baseline study to identify household characteristics; and a local community-based NGO, Shushilan, was responsible for implementing the project including capacity building for livelihoods. Importantly, a Project Implementation Committee (PIC) was formed with members from the beneficiary community to supervise the project when it was implemented – this brought a sense of ownership to the community.

In line with CDMP's strategic institutional approach, a wide range of partners including government agencies, NGOs, academia at the national and local level were engaged in the project. There was an effort to maximise the skills and resources of the specific partners in this multidimensional project that went beyond only building houses.

Key features of the project

The Bainpara resettlement project, named as a 'Disaster Resilient Habitat' (Ministry of Disaster Management and Relief (MoDMR), n.d.); Nahar et al., 2015), is a major undertaking of CDMP II with an investment of US$375,000 – a significant amount of money in the local context. Beyond provision of land and housing, the project includes a range of other facilities and services to make the newly established settlement habitable and sustainable. As mentioned by a former CDMP staff member, the MoDMR often cites this project as an example of 'building back better'. Some of the key features are listed below:

- Each beneficiary household was given ownership of a land parcel of 240 square metres on which the house was built. To avoid flooding, the entire settlement was raised by soil from excavation of ponds and drainage canals through a cash-for-work programme, where predominantly women from the community worked to earn an income while contributing towards building their settlement elevated above flood level.
- 58 detached single-storey houses were built on the independent plots and connected by a paved walkway (see Figure 2.7). Each house is of 12 square metres in area and consists of a single room on a raised earthen plinth. The structure of the house is supported on four prefabricated

Figure 2.7 A view from Bainpara 'Disaster Resilient Habitat' settlement. Note the solar panel on the house, the sanitary latrine to the left and the range of extensions. Also note the settlements in the background being established in vulnerable locations

reinforced concrete posts held together by a concrete tie beam above. The lean-to type roof is made of prefabricated ferro-cement panels and the walls are of brick. In the local context, most houses are generally built of less durable organic materials – bamboo, timber or thatch – and corrugated iron sheets, which means they are easily damaged in disasters. The houses built in the project are much sturdier and are a definite improvement – in that sense, they are comparatively disaster-resilient. It is reported that they are built to withstand a wind speed of up to 260 kilometres per hour. The materials used are also built to be resistant to the local saline environment (Nahar et al., 2015). During strong windstorms, such as Cyclone Mahasen in 2013, people from the surrounding villages took refuge in these houses (Nahar et al., 2015), and as one agency staff remarked, these houses were 'mini-cyclone shelters'.

- Each house included a rainwater tank for water supply. The storm surge from Cyclone Aila that inundated the area contaminated the local water bodies that were previously a source of drinking water. Additionally, in this coastal location, climate change-induced salinity intrusion is a problem, as discussed above in the section on the disaster profile of Bangladesh. Hence the rainwater tanks are essential to ensure drinking water supply to the community.

- A solar panel for generating electricity was provided to each house. The area is beyond reach of the national electricity grid – and previously the community, as other communities in the area, relied on kerosene lamps and other such sources of lighting. It is reported that the solar power enables children to study in the evening, among other such benefits (Nahar et al., 2015).
- A sanitary latrine was included with each house. Previously, such latrines were uncommon – open defecation or hanging latrines over water bodies was widely prevalent, resulting in the spread of gastro-intestinal diseases. The sanitary latrines now contribute to the health and hygiene of the community.
- Three freshwater ponds were provided in the settlement. These are used for a variety of purposes, including bathing and rearing fish and ducks. A solar-powered Pond Sand Filter (PSF) was installed next to one of the ponds to derive clean drinking water; other activities are restricted in this pond to keep it clean. The PSF water also benefits people from surrounding villages, not only the settlement's community. Cooperatives have been established to share the fish produced in the ponds.
- Community facilities were built, including a school, mosque, graveyard and market shed. Importantly, a cyclone shelter that provides safe refuge to the surrounding communities was built. During normal times this functions as a multi-purpose community building, including a pre-school.
- Trees were planted around the settlement as a 'green belt' (Nahar et al., 2015, p. 19) to serve as a protective buffer against the impact of cyclonic winds.
- Each household was given a grant of about US$160 as livelihood support. Training was provided by the NGO, Shushilan, to utilise the funds to generate income through activities such as homestead gardening and rearing poultry and livestock. Since then, beneficiaries have invested in additional resources on their own, such as purchasing boats to sell produce in the wider region, including Khulna city.

Thus, the wide range of facilities and support provided to the Bainpara community enabled it to establish itself as a habitable settlement. It is a marked improvement from pre-cyclone living conditions; after the prolonged suffering endured after Cyclone Aila, this was a significantly beneficial turning point for the community.

Key success factors

This project offers potential for replication to address the housing needs of the landless, as well as vulnerable and low-income communities. Together with housing, the settlement includes a range of facilities for water supply and sanitation, solar energy, livelihoods, and a school and cyclone shelter. Planned with fishponds and vegetation integrated with the housing, the settlement has gone beyond being only a housing resettlement project. Being a community with pre-existing linkages, resettling the community together allows cohesiveness. This planned settlement shows indications of a resilient community. Beneficiaries reported satisfaction with the project.

The housing has produced positive social and economic outcomes. From enhancing children's education to providing a safe place for home-based income-generating activities, the housing allows the community to flourish. The houses, built with brick and concrete, are well-constructed and durable, providing a level of safety from hazards such as cyclones and floods to an extent that the beneficiaries never had before.

While the house size is small at 12 square metres, the plot size of 240 square metres is comparatively large, allowing scope for extensions and home-gardens. Typically, Bangladeshi rural houses have a covered veranda in front, and most of the beneficiary households have added a veranda. Also, a rural homestead tends to have a cluster of small buildings, each accommodating a separate function (for a description of typical rural housing in Bangladesh see Ahmed, 2012). Thus, beyond the main house provided in the project, beneficiaries have invested in constructing additional buildings on their plots and many have built extensions. Beneficiaries have also widely established home gardens and fruit trees. The settlement does seem to be flourishing and has moved a step forward from its initial beginning.

Key challenges

While there are many successes of the project, the rural context of a remote coastal area also posed a number of challenges. There were several issues regarding the housing, despite being built to a good standard: to ensure quality, most of the concrete house components were prefabricated by HBRI in Dhaka and then transported to the project site, adding to the cost and time taken to complete the project. The prefabricated concrete posts and beams and ferro-cement roofing channels were transported by truck and then boat, and even by foot to the remote area – a tedious and cumbersome process. Transport of materials to remote areas, in addition to the extra cost, includes embodied energy costs. There is also the possibility of breakage and loss during transporting, which happened in this project. It was reported though that repairs were easily made with cement, but again, that added to the cost and inconvenience.

Producing these building components locally was not explored, which could have been a basis for introduction of new construction technologies to the region and wider replication. Thus, the opportunity to assess whether the use of prefabricated building components has potential for replication in rural areas was missed. Perhaps it would require an additional replication programme, but the feasibility of that remains uncertain because of the missed opportunity.

Because these construction technologies were not local, long-term maintenance might present problems in this remote location due to lack of local workmanship and availability of materials – for example, good quality cement, aggregates, steel rebar and wire mesh.

Community consultation and participation was lacking in this project and by and large the planning, design and implementation was controlled by CDMP and its partner organisations. There was an attempt to include local participation by forming the PIC with community members, as mentioned above when discussing the project's features. However, by that time the project planning and design of the settlement and housing had already been completed.

The 12-square metre one-room houses are very small and cannot accommodate the typically large families in Bangladesh. As mentioned above when discussing the success factors, most beneficiaries have built verandas or additional rooms with local materials on one or two sides of the house for extra space, and additional buildings are made of locally available and affordable materials. While this demonstrates the initiative of the beneficiaries to invest towards their own improvement, it should be noted that these additions would not be able to withstand the ferocious cyclones that frequently assail the area. It is fortunate though that the main house is sturdier and can function as a core shelter. It was also noted that asbestos roofing sheets were used widely for roofing of the verandas, which poses a serious health risk. Thus, there is scope to go beyond the original project and provide long-term support to the community to build awareness and further improve its resilience.

Although a wide range of facilities were provided, the settlement lacked medical services. The nearest clinic was in the town of Chalna, about 30 kilometres away. The remoteness of the settlement's location and lack of adequate transport services makes it difficult to access the service and presents a serious problem during emergencies. This points to a broader issue: remote coastal areas such as this in Bangladesh widely lack basic services. This is a national development challenge not only specific to this settlement.

Despite the achievements of the project, it remains as a small island in a vast ocean of need; 58 is a small number of houses relative to the thousands of people in the area who live a vulnerable and marginal existence. In that sense, as commented by an expert during an interview, 'The Bainpara settlement is disconnected from the surrounding villages.' From the settlement, it can be seen that extensive informal settlements have been established in vulnerable locations including outside the embankment (see Figure 2.7), evincing a situation similar to before Cyclone Aila struck. Thus, while the Bainpara settlement will be in a position of relative safety during a future cyclone, the fate of its neighbouring communities will be uncertain. Admittedly, a project such as this has a limited scope, but it brings to the forefront the importance of addressing the disaster resilience needs of the wider population of coastal areas of Bangladesh.

Limitations of CDMP

CDMP was a unique and extensive programme and has an impressive record of achievements and activities. In a developing country context such as Bangladesh, there are many constraints to operationalise and implement such an ambitious programme, and thus the programme is remarkable in how much it accomplished. Yet, experts and others associated with the programme pointed out some of its limitations, as discussed below, reflecting the broader institutional challenges discussed above in the section on policy instruments and challenges.

Lack of follow-up

A key issue was that there was no phase-out or exit strategy – in that sense, the efforts of CDMP were not sustained beyond the project's time frame. Once the project was completed there was little follow-up or maintenance of the programme's inputs. Although it was a comprehensive programme, it could not be mainstreamed over the long run into national disaster resilience policies, programmes and practices. As observed by an expert, 'There were too many interventions, but no follow-up.' All the rhetoric about long-term disaster risk reduction seemed to have ceased after the programme ended, at least in the sense that it had been advocated within the programme. As another expert noted, 'There were many good ideas [such as the risk reduction paradigm shift], but the government failed when trying to implement them and then could not hold on to the ideas.'

Much of the work, for example the Bainpara project discussed above, was not scaled up and replicated on a wider national basis. Many of these projects were conceived as pilot projects, which are generally intended to demonstrate the efficacy of certain interventions and to generate lessons for future application. However, the lessons from the CDMP projects were not utilised and the pilot projects in most cases remain isolated examples and were not replicated. Thus, the government ended up in only short-term service provision without a long-term view.

It should, however, be acknowledged that some related initiatives have been undertaken after CDMP was completed. For example, the three-year (2018–2021) National Resilience Programme (NRP) (National Resilience Programme, n.d.) is a similar partnership that has been developed between MoDMR and UNDP, along with other national stakeholders, and funded by some of the donors that supported CDMP (DFID and SIDA). However, the shorter time frame and funding (US$ 12 million) is small compared to either of the two CDMP phases and thus the scope is more limited. Nonetheless, to some extent the NRP represents an effort towards continuity of the CDMP legacy. Perhaps some of the lessons gained from CDMP might inform the actions of the NRP and other such initiatives.

Lack of sustainability of inputs

As discussed in the section on CDMP's key features, a wide range of capacity building initiatives were supported and undertaken. But again, because of the lack of post-programme follow-up, the long-term impacts of these remain uncertain. For example, about 40 government personnel were sent for training on disaster resilience at a high cost to Australia, but currently they cannot be tracked to understand whether and how they continue to apply the training in their work; because of frequent staff turnover, they have become scattered. A similar lack of sustainability of the capacity building initiatives is evident in other domains. A number of short-term project-based positions were created, but most of the people employed in these positions lost their jobs and went elsewhere, taking with them the capacity developed through their involvement in the programme.

An extensive array of knowledge products was produced under CDMP, but their current status remains elusive to most disaster resilience actors and stakeholders. Some of the key products – such as CRA training manuals, a detailed disaster risk atlas and earthquake risk assessments – were not maintained or upgraded and were not applied after the programme ended. There was hardly any capacity within the government to utilise some of the knowledge products because of the technical expertise required (such as the GIS-based risk assessment tools). A high-tech knowledge management system was developed at a high cost, employing expert consultants, but because of the lack of an adequate exit strategy, this was not transferred to the government by building the required capacity. A high-quality website was created, but subsequently, the government (represented by MoDMR) was reluctant to pay for the licence and it was discontinued. The repository and dissemination mechanisms for the knowledge products that were established during CDMP's term – such as the e-library, Disaster Management Information Centre (DMIC) and the Disaster Management Information Network (DMIN) – all ceased operation after the project ended. A significant amount of funding from donor agencies was invested in the programme, but there was little effort or exploration on how the investment could be sustained over the long term.

Reliance on external funding

A critical issue is funding. The large bulk of funding for both CDMP I and CDMP II were from international donor agencies with minimal financial contribution by the government. From the government's point of view (in this case MoDMR), it does not have the funds to scale up, replicate or mainstream the initiatives undertaken through CDMP. In this context, as remarked by an expert, 'The government is only interested in donor funds.'

There are three main aspects to this issue. Firstly, there is an instrumental approach of the government that programmes and projects can be implemented only if there is external funding available, indicating a reliance on such funds which is not sustainable. As one expert commented, 'UNDP has spoilt the government.' Secondly, unless such funding is incorporated into the government's revenue – such as through provision of employment for public sector personnel – the government does not seem interested in wider national development and resilience. Thirdly, the government appears to be more attracted to large projects, especially when there is infrastructure construction involved (cyclone shelters, culverts, etc. in the case of disaster resilience projects), because of the opportunities for kickbacks; small-scale grassroots projects, though they are highly relevant for Bangladesh, have little appeal.

Inadequate capacity

The main focus in CDMP I was on policy reform and building capacity, where only a few problems were encountered such as 'inter-ministerial multi-sectoral coordination' (Russell et al., 2009, p. 21) – that is in mainstreaming the paradigm shift from responding only to the impacts of disasters to addressing the causal factors behind disasters. However, the next phase, CDMP II, involved actual interventions and projects, and this is where the lack of capacity within the government became evident. There were only three to four staff members employed to run CDMP II and it became very difficult, if not impossible at times, to monitor and supervise the projects effectively. In this sense, even though the government claimed ownership of the programme, it largely failed in managing it. It was treated as a separate programme and not integrated into the overall process of the ministry's operation.

Some of the experts who were interviewed suggested that a 'more focused' approach was required – the projects should have been 'knowledge-based' or 'value-based', derived from a cost–benefit analysis – but it seems that the projects were largely 'needs-based', that is, reactive in nature and usually in response to disaster impacts. Thus, ironically, the government somewhat fell back to the previous mode that the paradigm shift initiative sought to challenge. Because of the lack of follow-up as mentioned above, there was no further capacity strengthening after the project was completed, especially at the grassroots level.

Conclusion

The deltaic context of Bangladesh, comprised predominantly of low-lying floodplains and with a long southern coastline, experiences recurrent floods and cyclones – two of the main hazards that have catalysed institutional initiatives for disaster resilience. Human mortality resulting from cyclones has been reduced significantly over the past few years through a combination of structural and non-structural measures. However, economic losses incurred from disasters continue to mount – especially from floods – because urbanisation and population growth places more and more assets at risk, and also due to climate change causing more extensive and frequent floods. Bangladesh is beset by a range of other natural hazards, keeping institutions busy throughout the year: during the monsoon season, there is too much water from floods, and in the dry season there is water scarcity causing drought. On top of being affected by natural hazards, human-induced hazards such as building collapses and transport accidents place great demands on the limited reserves of institutions. Thus, disaster resilience will need to be an institutional priority for the foreseeable future.

Despite recent economic growth, Bangladesh continues to have a high prevalence of poverty, aggravated by the frequent and recurrent disasters that are increasingly amplified by climate change. The challenges of implementing disaster resilience initiatives in such a context are hence severe. The flipside of economic growth – rapid and largely unplanned urbanisation – is widespread, not only in the main cities, but also the multitude of smaller cities and peri-urban areas. A serious risk of earthquakes exists, making such unregulated urban development potentially vulnerable to catastrophic impacts. Additionally, economic development exacts a harsh toll on the environment from an incessant demand on natural resources and pollution from industrial production processes.

Because of the long history of disasters and the associated human tragedies, Bangladesh has been at the forefront of disaster resilience initiatives for a long time; this has been led by a dedicated ministry, MoDMR, which is quite unique to the South Asian region. There are extensive policy instruments in place, which is why a large initiative such as CDMP was embraced from the early 2000s. A well-funded and carefully developed programme, CDMP spearheaded the 'paradigm shift' to focus on disaster resilience beyond only post-disaster response; CDMP profoundly impacted the thinking in all circles in Bangladesh that deal with disasters. Implemented over more than a decade in two phases, the initial policy-focused phase led to a more robust phase of implementation of projects. CDMP's portfolio consists of an impressive record of achievements. The Bainpara 'Disaster Resilient Habitat' project discussed in this chapter is only an indicative example of CDMP's track record, but it demonstrates how a comprehensive and integrated approach to post-disaster housing resettlement can create a habitable and sustainable settlement, despite various limitations.

The key problems arose after completion of the programme. In spite of the extensive investments towards addressing the disaster risks in Bangladesh, a follow-up strategy barely existed. Not being nurtured beyond the project time frame, the investments seem to have achieved mostly short-term outcomes, at least for the government's disaster resilience system. Perhaps the key issue is of a recalcitrant nature – the instrumental reliance on external funds influencing governmental decision-making processes, diminishing any initiative for independent action and sustaining investments, and to merely act as service providers as long as funds last. Bangladesh has in many ways been a leader in the disaster resilience field, but to graduate to the next level a more proactive and innovative approach driving institutional initiatives will be required.

References

Ahmad, S.A., Khan, M.H. and Haque, M. (2018) 'Arsenic Contamination in Groundwater in Bangladesh: Implications and Challenges for Healthcare Policy'. *Risk Management and Healthcare Policy*, 11, pp. 251–261.

Ahmed, B. (2017) *Community Vulnerability to Landslides in Bangladesh* (Doctoral dissertation). London: University College London.

Ahmed, I. (2000) 'Post-Flood Investigations 1998 and an Action Agenda for Rural Housing in Bangladesh' in Seraj, S.M., Hodgson, R. and Choudhury, J.R. (eds.) *Affordable Village Building Technologies*. Dhaka: Bangladesh University of Engineering & Technology and Exeter: University of Exeter, pp. 135–153.

Ahmed, I. (2007) *Emergency Telecommunications and Early Warning Systems for Disaster Preparedness in Chittagong, Bangladesh*. Geneva: International Telecommunications Union (ITU).

Ahmed, I. (2012) 'The Courtyard in Rural Homesteads of Bangladesh'. *Vernacular Architecture*, 43(1), pp. 47–57.

Ahmed, I. (2014) 'Factors in Building Resilience in Urban Slums of Dhaka, Bangladesh'. *Procedia Economics and Finance*, 18, pp. 745–753.

Ahmed, I. (2016) 'Developing Disaster Resilient Rural Housing in Coastal Bangladesh'. *Sustainable Futures Conference on Architecture and Construction in the Global South*, Nairobi, pp. 186–196.

Ahmed, I. (2019) 'The National Plan for Disaster Management of Bangladesh: Gap between Production and Promulgation'. *International Journal of Disaster Risk Reduction*, 37, pp. 1–8.

Ahmed, I., Gajendran, T., Brewer, G., Maund, K., von Meding, J., Kabir, H., Faruk, M., Shrestha, H.D. and Sitoula, N.R. (2018) *Understanding the Opportunities and Challenges of Compliance to Safe Building Codes for Disaster Resilience in South Asia* (technical report). Kobe: Asia Pacific Network for Global Change Research (APN).

Ahmed, I. and Johnson, G. (2013) 'Urban Safety and Poverty in Dhaka, Bangladesh: Understanding the Structural and Institutional Linkages'. *Australian Planner*, 51(3), pp. 272–280.

Al Jazeera. (2019) *Worst Floods in Years 'Submerge' Bangladesh Villages*. Available at: www.aljazeera.com/news/2019/07/worst-floods-years-submerge-bangladesh-villages-190719083053518.html (accessed 22 July 2019).

Alam, E. and Ray-Bennett, N.S. (2019) *Enhancing Local Level Disaster Risk Governance in Bangladesh* (PowerPoint presentation). Chittagong: University of Chittagong.

Alamgir, M., Shahid, S., Hazirak, M.K., Nashrrullah, S., Harun, S.B. and Shamsudin, S. (2015) 'Analysis of Meteorological Drought Pattern during Different Climatic and Cropping Seasons in Bangladesh'. *Journal of the American Water Resources Association*, 51(3), pp. 794–806.

Angeles, G., Lance, P., Barden-O'Fallon, J., Islam, N., Mahbub, A.Q.M. and Nazem, N.I. (2009) 'The 2005 Census and Mapping of Slums in Bangladesh: Design, Select Results and Application'. *International Journal of Health Geographics*, 8, pp. 1–19.

Asian Development Bank (ADB). (2013) *Bangladesh: Jamuna–Meghna River Erosion Mitigation Project*. Available at: www.adb.org/projects/34038-013/main (accessed 7 October 2019).

Bala, S.K. (2018) *Our Wetlands and a Sustainable Urban Future*. Daily Star. Available at: www.thedailystar.net/supplements/rethinking-urban-spaces/our-wetlands-and-sustainable-urban-future-1537972 (accessed 25 July 2019).

Bangladesh National Parliament. (2012) *Bangladesh Gazette*. Dhaka: Government of the People's Republic of Bangladesh.

Bangladesh Passenger Welfare Organisation. (2019) *Annual Road Accident Observation Report 2018*. Dhaka: Bangladesh Passenger Welfare Organisation.

Bangladesh Planning Commission (BPC). (2012) *The Millennium Development Goals Bangladesh Progress Report 2011*. Dhaka: BPC.

Bangladesh Planning Commission (BPC). (2018) *Sustainable Development Goals Bangladesh Progress Report 2018*. Dhaka: BPC.

Bangladesh Red Crescent Society (BDRCS). (2011) *Early Response Recovery for Cyclone Aila Affected most Vulnerable People of Khulna District, Bangladesh* (Project Proposal). Dhaka: BDRCS.

Bangladesh Red Crescent Society (BDRCS). (2019) *Cyclone Preparedness Programme (CPP)*. Available at: www.bdrcs.org/cyclone-preparedness-programme-cpp (accessed 18 July 2019).

Bangladesh Water Development Board (BWDB). (2014) *Bangladesh: Flood and Riverbank Erosion Risk Management Investment Program*. Dhaka: BWDB.

Banks, N., Roy, M. and Hulme, D. (2011) 'Neglecting the Urban Poor in Bangladesh: Research, Policy and Action in the Context of Climate Change'. *Environment and Urbanization*, 23(2), pp. 487–502.

BBC News. (2017) *Deadly Cyclone Mora Hits Bangladesh with High Winds and Rain*. Available at: www.bbc.com/news/world-asia-40089409 (accessed 30 September 2019).

BBC News. (2019) *Cyclone Fani: More than 1,000 Bangladesh Homes Destroyed*. Available at: www.bbc.com/news/world-asia-48165795 (accessed 30 September 2019).

BDNews24. (2014) *Hailstorm Lashes Dhaka*. Available at: http://bdnews24.com/bangladesh/2014/04/04/hailstorm-lashes-dhaka (accessed 31 July 2019).

Brammer, H. (1990) 'Floods in Bangladesh: Geographical Background to the 1987 and 1988 Floods'. *Geographical Journal*, 156(1), pp. 12–22.

Briney, A. (2018) *Countries of Asia by Area*. ThoughtCo. Available at: www.thoughtco.com/countries-of-asia-by-area-1434341 (accessed 30 September 2019).

Byron, R.K. and Parvez, S. (2019) *Inequality at All-Time High*. Daily Star. Available at: www.thedailystar.net/frontpage/news/inequality-all-time-high-1745569 (accessed 22 July 2019).

Center for Environmental and Geographic Information Services (CEGIS). (2009) *Report on Cyclone Shelter Information for Management of Tsunami and Cyclone Preparedness Annex C: Catchments Area and Evacuation Route Mapping*. Dhaka: CEGIS.

Center for Environmental and Geographic Information Services (CEGIS). (2013) *Vulnerability to Climate Induced Drought: Scenario & Impacts*. Dhaka: Comprehensive Disaster Management Programme (CDMP II) and Ministry of Disaster Management and Relief.

Center for Excellence in Disaster Management & Humanitarian Assistance (CEDMHA). (2017) *Bangladesh Disaster Management Reference Handbook*. Available at: www.cfe-dmha.org/(accessed 6 October 2019).

Central Intelligence Agency (CIA). (2019) *The World Fact Book*. Available at: www.cia.gov/library/publications/resources/the-world-factbook/geos/bg.html (accessed 30 July 2019).

Charlesworth, E. and Ahmed, I. (2015) *Sustainable Housing Reconstruction: Designing Resilient Housing After Natural Disasters*. Abingdon (UK): Routledge.

Daily Star. (2016) *Bangladesh Wakes Up to Tremors*. Available at: www.thedailystar.net/country/bangladesh-wakes-tremors-196783 (accessed 10 October 2019).

Daily Star. (2018) *Shun Burnt Bricks to Save Top Soil*. Available at: www.thedailystar.net/backpage/news/manufacturing-burnt-bricks-100m-tonnes-top-soil-ruined-year-1636384 (accessed 25 July 2019).

Daily Star. (2019) *Stop Illegal Sand Extraction from Rivers or Quit*. Available at: www.thedailystar.net/country/news/stop-illegal-sand-extraction-rivers-or-quit-1758319 (accessed 25 July 2019).

Das, M.K. and Karmakar, S. (2015) 'Urban Heat Island Assessment for a Tropical Urban Air-shed in Bangladesh'. *9th International Conference on Urban Climate*, Toulouse (France).

Dasgupta, S., Huq, M., Khan, Z.H., Ahmed, M.M.A., Mukherjee, N., Khan, M.F. and Pandey, K. (2011) *Cyclones in a Changing Climate: The Case of Bangladesh*. Washington, DC: World Bank.

Dey, N.C., Alam, M.S., Sajjan, A.K., Bhuiyan, M.A., Ghose, L., Ibaraki, Y. and Karim, F. (2011) 'Assessing Environmental and Health Impact of Drought in the Northwest Bangladesh'. *Journal of Environmental Science and Natural Resources*, 4(2), pp. 89–97.

Dhaka Tribune. (2017) *To Cut Brick Kiln Pollution, Bangladesh Constructs New Building Materials*. Available at: www.dhakatribune.com/bangladesh/environment/2017/11/09/brick-kiln-pollution-bangladesh-new-building-materials/(accessed 25 July 2019).

Dhaka Tribune. (2018) *Illegal Sand Mining Continues on Kharkharia Floodplain*. Available at: www.dhakatribune.com/bangladesh/environment/2018/03/30/illegal-sand-mining-kharkharia-floodplain (accessed 25 July 2019).

Dhaka Tribune. (2019a) *Sand Extraction from Khutakhali Falls Threatens Lives in Naikhongchhari*. Available at: www.dhakatribune.com/opinion/special/2019/03/28/sand-extraction-from-khutakhali-falls-threatens-lives-in-naikhongchhari (accessed 25 July 2019).

Dhaka Tribune. (2019b) *Fire Incidents in Bangladesh Triple over 22 Years*. Available at: www.dhakatribune.com/bangladesh/nation/2019/02/26/fire-incidents-in-bangladesh-triple-over-22-years (accessed 1 August 2019).

Displacement Solutions. (2012) *Climate Displacement in Bangladesh the Need for Urgent housing, Land and Property (HLP) Rights Solutions*. Portsea (Australia): Displacement Solutions.

Dolce, C. (2014) *The Deadliest Tornado in the World*. The Weather Channel. Available at: https://weather.com/storms/tornado/news/deadliest-tornado-world-bangladesh-20140416#2 (accessed 31 July 2019).

Emergency Events Database (EMDAT). (2015) *Bangladesh Disaster & Risk Profile*. Available at: www.preventionweb.net/countries/bgd/data/(accessed 30 July 2019).

Finch, J.D. and Dewan, A.M. (2007) *Tornadoes in Bangladesh and East India*. National Weather Service (USA). Available at: http://bangladeshtornadoes.org/climo/btorcli0.htm (accessed 31 July 2019).

George, S. (2017) *A Third of Bangladesh under Water as Flood Devastation Widens*. CNN. Available at: https://edition.cnn.com/2017/09/01/asia/bangladesh-south-asia-floods/index.html (accessed 22 July 2019).

Glennon, R. (2017) *The Unfolding Tragedy of Climate Change in Bangladesh*. Scientific American. Available at: https://blogs.scientificamerican.com/guest-blog/the-unfolding-tragedy-of-climate-change-in-bangladesh/(accessed 1 August 2019).

Global Facility for Disaster Risk Reduction (GFDRR). (n.d.) *Tsunami*. Available at: http://thinkhazard.org/en/report/23-bangladesh/TS (accessed 01 July 2019).

Government of Bangladesh (GoB). (2008) *Cyclone Sidr in Bangladesh: Damage, Loss, and Needs Assessment for Disaster Recovery and Reconstruction*. Dhaka: GoB.

Green Bangla. (2013) *River Bank Erosion Map of Bangladesh*. Available at: http://geobird104.blogspot.com/2013/03/river-bank-erosion-map-of-bangladesh.html (accessed 31 July 2019).

The Guardian. (2019) *Rohingya Crisis: Bangladesh Says It Will Not Accept Any More Myanmar Refugees*. Available at: www.theguardian.com/world/2019/mar/01/rohingya-crisis-bangladesh-says-it-will-not-accept-any-more-myanmar-refugees (accessed 1 August 2019).

Halder, S.R. (2010) *CDMP 2004-2009: Laying the Foundations for Comprehensive Disaster Management*. Dhaka: Comprehensive Disaster Management Programme.

Halim, S. and Chowdhury, K. (2015) 'The Land Problem in the Chittagong Hill Tracts: A Human Rights Anatomy' in National Human Rights Commission (2016) *JAMAKON Year Book, 2015*. Dhaka: National Human Rights Commission, pp. 36–80.

Haq, T. (2013) 'A Violent Threat to Conservation of Wetlands and the Existing Laws of Bangladesh: A Critical Analysis'. *Philosophy and Progress*, LIII–LIV, pp. 113–126.

Haque, C.E. and Uddin, M.S. (2013) *Disaster Management Discourse in Bangladesh: a Shift from Post-Event Response to the Preparedness and Mitigation Approach Through Institutional Partnerships*. IntechOpen. Available at: www.intechopen.com/books/approaches-to-disaster-management-examining-the-implications-of-hazards-emergencies-and-disasters/disaster-management-discourse-in-bangladesh-a-shift-from-post-event-response-to-the-preparedness-and (accessed 5 October 2019).

Haque, U., Hashizume, M., Kolivras, K.N., Overgaard, H.J., Das, B. and Yamamoto, T. (2012) 'Reduced Death Rates from Cyclones in Bangladesh: What More Needs to Be Done?'. *Bulletin of the World Health Organization*, 90, pp. 150–156.

Hashim, S.M. (2019) *A Quarter of Dhaka's Wetlands Gone*. Daily Star. Available at: www.thedailystar.net/opinion/no-frills/news/quarter-dhakas-wetlands-gone-1739869 (accessed 25 July 2019).

Hasnat, M.A. (2018) *Cold Freezes Normal Life across Bangladesh*. Dhaka Tribune. Available at: www.dhakatribune.com/bangladesh/nation/2018/01/05/cold-spell-wave-across-bangladesh (accessed 31 July 2019).

Haustein, K., Uhe, P., Rimi, R., Islam, A.S. and Otto, F.E.L. (2017) *Changing Frequency of Flooding in Bangladesh: Is the Wettest Place on Earth getting Wetter?*. ADS. Available at: http://adsabs.harvard.edu/abs/2017AGUFM.A33E2410H (accessed 22 July 2019).

Hindustan Times. (2010) 'Experts Warn of Earthquakes in Bangladesh'. *Hindustan Times* (daily), 17 September 2010.

Hong Kong Red Cross. (2009) *Bangladesh Cyclone Aila: Work Report 1*. Available at: www.redcross.org/hk/en/disasterelief_prepared/international_projects/ipappeal_update_detail.html?year=2009&id=409 (accessed 10 October 2013).

Hosen, M.S. and Jubayer, A. (2016) 'Chronological History and Destruction Pattern of Tornados in Bangladesh'. *American Journal of Environmental Protection*, 5(4), pp. 71–81.

Housing and Building Research Institute (HBRI). (2015) *Bangladesh National Building Code (Final Draft)*. Dhaka: HBRI.

Human Rights Watch. (2016) *Nepotism and Neglect: The Failing Response to Arsenic in the Drinking Water of Bangladesh's Rural Poor*. Available at: www.hrw.org/report/2016/04/06/nepotism-and-neglect/failing-response-arsenic-drinking-water-bangladeshs-rural (accessed 31 July 2019).

Iaminweb. (2010). *Earthquake Scare Dhaka Residents Fear Dangerous Strike*. Available at: http://iaminweb.com/miscellaneous.php (accessed 10 April 2014).

Inform. (2019) *Inform Report 2019: Shared Evidence for Managing Crises and Disasters*. Available at: https://reliefweb.int/report/world/inform-report-2019-shared-evidence-managing-crises-and-disasters (accessed 15 October 2019).

Institute of Economics and Peace. (2019) *Global Peace Index 2019 Measuring Peace in a Complex World*. Available at: https://reliefweb.int/report/world/global-peace-index-2019 (accessed 15 October 2019).

Intergovernmental Panel on Climate Change (IPCC). (2012) *Managing the Risks of Extreme Events and Disasters to Advance Climate Change Adaptation. A Special Report of Working Groups I and II of the Intergovernmental Panel on Climate Change*. Cambridge: Cambridge University Press.

International Federation of Red Cross and Red Crescent Societies. (2018) *World Disasters Report: Leaving No One Behind*. Available at: www.ifrc.org/wdr2018 (accessed 6 October 2019).

Islam, M.F. and Rashid, A.B. (2012) 'Riverbank Erosion Displaces in Bangladesh: Need for Institutional Response and Policy Intervention'. *Bangladesh Journal of Bioethics*, 2(2), pp. 4–19.

Islam, M.M. (2013) *Country Report: People's Republic of Bangladesh*. Kobe: Asian Disaster Reduction Center.

Islam, M.S. (2018) *Lightning Kills Two, Crops Damaged by Hailstones in Rangpur, 1 Dead in Sylhet*. Dhaka Tribune. Available at: www.dhakatribune.com/bangladesh/nation/2018/03/31/lightning-kills-two-crops-damaged-hailstones-rangpur-1-dead-sylhet (accessed 31 July 2019).

Islam, M.S. and Islam, F. (2019) *Arsenic Contamination in Groundwater in Bangladesh: An Environmental and Social Disaster*. IWA Publishing. Available at: www.iwapublishing.com/news/arsenic-contamination-groundwater-bangladesh-environmental-and-social-disaster (accessed 31 July 2019).

Islam, R., Islam, M.N. and Islam, M.N. (2016) 'Earthquake Risks in Bangladesh: Causes, Vulnerability, Preparedness and Strategies for Mitigation'. *ARPN Journal of Earth Science*, 5(2), pp. 75–90.

Islamic Relief. (2014) *Still Feeling the Toll of Cyclone Aila*. Available at: www.islamic-relief.org.uk/cyclone-aila-five-years/(accessed 11 October 2019).

Kabir, M.H., Sato, M., Habbiba, U. and Yousuf, T.B. (2018) 'Assessment of Urban Disaster Resilience in Dhaka North City Corporation (DNCC), Bangladesh'. *Procedia Engineering*, 212, pp. 1107–1114.

Kafle, S.K. (2017) *Disaster Risk Reduction: Case Studies from Asia*. Kathmandu: Disaster and Climate Change Study Centre.

Kamal, A.M., Shamsudduha, M., Ahmed, B., Hassan, S.K., Islam, M.S., Kelman, I. and Fordham, M. (2018) 'Resilience to Flash Floods in Wetland Communities of Northeastern Bangladesh'. *International Journal of Disaster Risk Reduction*, 31, pp. 478–488.

Khan, A.N.M.M. (2010) *Impact of Climate Change on the Livelihood of the Urban Poor: A Case Study of Dhaka City* (Master's thesis). Dhaka: North South University.

Khatun, R., Gossami, G.C., Akter, S., Barman, M.C. and Paul, G.C. (2017) 'Impact of Tropical Cyclone Aila along the Coast of Bangladesh'. *International Journal of Scientific & Engineering Research*, 8(12), pp. 1592–1599.

Kreft, S., Eckstein, D., Jughans, L., Kerestan, C. and Hagen, U. (2014) *Global Climate Risk Index 2015: Who Suffers Most from Extreme Weather Events? Weather-related Loss Events in 2013 and 1994 to 2013*. Bonn: Germanwatch e.V.

Lovelle, M. (2016) *India, Bangladesh and the Farakka Barrage*. Future Directions International. Available at: www.futuredirections.org.au/publication/india-bangladesh-farakka-barrage/(accessed 15 October 2019).

Mallick, F. and Rahman, A. (2013) 'Cyclone and Tornado Risk and Reduction Approaches in Bangladesh' in Shaw, R., Mallick, F. and Islam, A. (eds.) *Disaster Risk Reduction Approaches in Bangladesh*. Tokyo: Springer, pp. 91–102.

Miller, B. (2016) *Bangladesh: Hidden Fault Could Trigger Major Quake*. CNN. Available at: https://edition.cnn.com/2016/07/21/weather/bangladesh-earthquake-threat/index.html (accessed 30 July 2019).

Minar, M.H., Hossain, M.B. and Shamsuddin, M.D. (2013) 'Climate Change and Coastal Zone of Bangladesh: Vulnerability, Resilience and Adaptability'. *Middle-East Journal of Scientific Research*, 13(1), pp. 114–120.

Ministry of Disaster Management and Relief (MoDMR). (2015) *Comprehensive Disaster Management Programme Phase II (Project Completion Report)*. Dhaka: MoDMR.

Ministry of Disaster Management and Relief (MoDMR). (n.d.) *Disaster Resilient Habitat, Bangladesh: An Alternative Comprehensive Approach to Disaster Risk Reduction and Climate Change Adaptation*. Dhaka: MoDMR.

Ministry of Environment and Forest (MOEF). (2006) *Impact of Sea-Level Rise on Land Use Suitability and Adaptation Options: Coastal Land Zoning in the Southwest*. Dhaka: Ministry of Environment and Forest.

Ministry of Food and Disaster Management (MoFDM). (n.d.) *Comprehensive Disaster Management Bangladesh Experience*. Dhaka: MoFDM.

Misachi. (2017) *Where Is the Largest Delta in the World?*. World Atlas. Available at: www.worldatlas.com/articles/which-is-the-largest-delta-in-the-world.html (accessed 11 October 2019).

Miyaji, M., Okazaki, K. and Ochiai, C. (2017) 'A Study on the Use of Cyclone Shelters in Bangladesh'. *Journal of Architecture and Planning*, 82(737), pp. 1871–1880.

Myers, J. (2016) *These Are the World's 10 Fastest Growing Megacities*. World Economic Forum. Available at: www.weforum.org/agenda/2016/11/the-10-fastest-growing-megacities-in-the-world/(accessed 2 October 2019).

Nahar, N.D., Ahmed, M., Osman, S., Quayyum, M.A., Khan, T.I., Parsons, L., Haarsaker, C.T. and Wijesinha, H. (2015) *Resilient Shelter Resilient Community: A Comprehensive Alternative Approach to Disaster Risk Reduction and Climate Change Adaptation*. Dhaka: Ministry of Disaster Management and Relief.

National Institute of Population Research and Training (NIPORT), Mitra and Associates, and ICF International. (2016) *Bangladesh Demographic and Health Survey 2014*. Dhaka and Rockville (USA): NIPORT, Mitra and Associates, and ICF International.

National Resilience Programme. (n.d.) www.nrpbd.org/index.php (accessed 14 September 2019).

New Age. (2011) *Major Quake to Bring Catastrophe in Dhaka*. Available at: www.newagebd.com/detail.php?date=2012-04-13&nid=7126 (accessed April 2014).

Parvin, G.A., Shimi, A.C., Shaw, R. and Biswas, C. (2016) 'Flood in a Changing Climate: The Impact on Livelihood and How the Rural Poor Cope in Bangladesh'. *Climate*, 4(4), pp. 60–75.

Paul, S.K., Hossain, M.N. and Ray, S.K. (2013) '"Monga" in Northern Region of Bangladesh: A Study on People's Survival Strategies and Coping Capacities'. *Rajshahi University Journal of Life & Earth and Agricultural Sciences*, 41, pp. 41–56.

Policy, Programme & Partnership Development Unit (PPPDU). (2009) *Comprehensive Disaster Management Programme Quarterly Progress Report*. Dhaka: Ministry of Food and Disaster Management.

PreventionWeb. (2014) *Bangladesh Disaster & Risk Profile*. Available at: www.preventionweb.net/countries/bgd/data/ (accessed 30 July 2019).

Rabbani, M.G. (2010) *Climate Change Vulnerabilities for Urban Areas in Bangladesh: Dhaka as a Case*. Dhaka: Bangladesh Centre of Advanced Studies.

Rahman, M.R. (2010) 'Impact of Riverbank Erosion Hazard in the Jamuna Floodplain Areas in Bangladesh'. *Journal of Science Foundation*, 8(1–2), pp. 55–65.

Rahman, M.R. and Lateh, H. (2016) 'Meteorological Drought in Bangladesh: Assessing, Analysing and Hazard Mapping Using SPI, GIS and Monthly Rainfall Data'. *Environmental Earth Sciences*, 75(12), pp. 1026–1046.

Rashid, H. (1993) *Preventing Flooding or Regulating Flood Levels?: Case Studies on Perception of Flood Alleviation in Bangladesh*. Available at: https://link.springer.com/article/10.1007/BF00596234 (accessed 22 July 2019).

Rashid, H.E. (1991) *Geography of Bangladesh*. Dhaka: University Press Limited.

Rezoane, Q.A. (2016) *A Study on Disaster Management Policy of Bangladesh: Institutional Capacity Enhancement* (postgraduate thesis). Dhaka: North South University.

Rokon, F. (2013) *Administrative Structure of Bangladesh*. SlideShare. Available at: www.slideshare.net/fahim002/administrative-structure-of-bangladesh (accessed 5 August 2019).

Rozario, R.R. and Uttom, S. (2014) *Bangladesh's Life-Threatening Chemical Hazards*. UCA News. Available at: www.ucanews.com/news/bangladeshs-life-threatening-chemical-hazards/72220 (accessed 1 August 2019).

Russell, N., Mahbub, A.Q.M., Khan, M.H. and Islam, N. (2009) *Bangladesh Comprehensive Disaster Management Programme Terminal Evaluation 2009*. Dhaka: Ministry of Food and Disaster Management.

Saari, M. and Rahman, S. (2003) 'Development of the Coastal Embankment System in Bangladesh' in Goudas, C., Katsiaris, G., May, V. and Karambas, T. (eds.) *Soft Shore Protection: Coastal Systems and Continental Margins, Vol 7*. Dordrecht: Springer, pp. 115–125.

Sachi, S.M. and Price, L.C. (2018) *Bangladesh's Air Pollution Problem Grows, Brick by Brick*. Pulitzer Center. Available at: https://pulitzercenter.org/reporting/bangladeshs-air-pollution-problem-grows-brick-brick (accessed 2 October 2019).

Sadik, M.S., Nakagawa, H., Rahman, R., Shaw, R., Kawaike, K. and Fujita, K. (2018) 'A Study on Cyclone Aila Recovery in Koyra, Bangladesh: Evaluating the Inclusiveness of Recovery with Respect to Predisaster Vulnerability Reduction'. *International Journal of Disaster Risk Science*, 9(1), pp. 28–43.

Sarker, P.K. (2009) '2.5 Lakh Buildings in Three Cities at Earthquake Risk'. *BD News*, 28 August 2009.

Sebastio, F. (2018) *Climate Change Is Threatening the Garment Industry*. GreenBiz. Available at: www.greenbiz.com/article/climate-change-threatening-garment-industry (accessed 30 July 2019).

Shafi, S.A. and Payne, G. (2007) *Local Partnerships for Urban Poverty Alleviation: Land Tenure Security and Land Administration in Bangladesh* (report). Dhaka: Local Government Engineering Department, United Nations Development Programme and UN-Habitat.

Shah, S. (1999) *Coping with Natural Disasters: The 1998 Floods in Bangladesh* (draft report). Washington, DC: World Bank.

Start Fund Bangladesh. (2019) *Start Disaster Summary Sheet: Riverbank Erosion*. Dhaka: Start Network.

Sunny, A.R. (2017) 'Impact of Oil Spill in the Bangladesh Sundarbans'. *International Journal of Fisheries and Aquatic Studies*, 5(5), pp. 365–368.

Trading Economics. (2019) *Bangladesh – Population in Largest City.* Available at: https://tradingeconomics.com/bangladesh/population-in-largest-city-wb-data.html (accessed 2 October 2019).

U.S. Geological Survey (USGS). (2012). *Magnitude 4.5 BANGLADESH, Sunday, March 18, 2012 at 02:56:10 UTC.* Available at: http://neic.usgs.gov/neis/bulletin/neic_b0008j9l_w.html (accessed 10 April 2014).

UN-Habitat. (2008) *State of the World's Cities 2008/2009.* London: Earthscan.

UN-Habitat. (2009) *Global Report on Human Settlements: Planning Sustainable Cities.* London: Earthscan.

UN-Habitat. (2010) *The State of Asian Cities 2010/11.* Fukuoka: UN-Habitat.

United Nations. (2018) *World Urbanization Prospects 2018.* Available at: https://population.un.org/wup/Country-Profiles/ (accessed 2 October 2019).

United Nations Development Programme (UNDP). (2019) *Bangladesh: Disaster Risk Reduction as Development.* Available at: www.undp.org/content/undp/en/home/librarypage/poverty-reduction/supporting_transformationalchange/Bangladesh-drr-casestudy-transformational-change.html (accessed 16 September 2019).

United Nations Development Programme (UNDP). (n.d.(a)) *Brief: Local Disaster Risk Reduction Fund (LDRRF): Comprehensive Disaster Management Programme Phase II.* Dhaka: UNDP.

United Nations Development Programme (UNDP). (n.d.(b)) *Project Factsheet: Comprehensive Disaster Management Programme Phase II.* Dhaka: UNDP.

United Nations Social and Economic Commission for the Asia Pacific (UNESCAP). (2015) *Disasters without Borders: Regional Resilience for Sustainable Development (Asia-Pacific Disaster Report 2015).* Bangkok: UNESCAP.

USAID. (2010) *Bangladesh.* Available at: www.land-links.org/country-profile/bangladesh/ (accessed 1 August 2019).

Uttom, S. and Rozario, R.R. (2019) *Rich-Poor Gap Increases in Bangladesh.* UCA News. Available at: www.ucanews.com/news/rich-poor-gap-increases-in-bangladesh/84334 (accessed 22 July 2019).

von Meding, J., Ahmed, I., Howard, A., Forino, G., Pal, I., Kabir, H., Dao, H. and Thai, H. (2018) *Critical Factors for Post-Disaster Educational Continuity in Urban Floods in South and Southeast Asia* (research report). Singapore: Save the Children.

Water and Sanitation for the Urban Poor (WSUP). (2007) *Water and Sanitation for the Urban Poor* (scoping report). London: WSUP.

Wikipedia. (2019) *Administrative Geography of Bangladesh.* Available at: https://en.wikipedia.org/wiki/Administrative_geography_of_Bangladesh (accessed 2 October 2019).

World Bank. (2007) *Dhaka: Improving Living Conditions for the Urban Poor.* Dhaka: World Bank Office.

World Bank. (2015) *Salinity Intrusion in a Changing Climate Scenario Will Hit Coastal Bangladesh Hard.* Available at: www.worldbank.org/en/news/feature/2015/02/17/salinity-intrusion-in-changing-climate-scenario-will-hit-coastal-bangladesh-hard (accessed 1 August 2019).

World Bank. (2019) *The World Bank in Bangladesh.* Available at: www.worldbank.org/en/country/bangladesh/overview (accessed 2 October 2019).

World Population Review. (2019a) *World Population by Country.* Available at: http://worldpopulationreview.com/ (accessed 30 September 2019).

World Population Review. (2019b) *Bangladesh Population 2019*. Available at: http://worldpopulationreview.com/countries/bangladesh-population/(accessed 30 September 2019).

WorldData. (n.d.) *Tsunamis in Bangladesh*. Available at: www.worlddata.info/asia/bangladesh/tsunamis.php (accessed 1 August 2019).

Worldometers. (2019) *Bangladesh Population*. Available at: www.worldometers.info/world-population/bangladesh-population/(accessed 25 July 2019).

Zahid, A. (2017) *Establishment of Monitoring Network and Mathematical Model to Assess Saline Water Intrusion in Groundwater in the Coastal Area of Bangladesh* (poster). Dhaka: Bangladesh Water Development Board.

The 1994 glacial lake outburst

Climate change adaptation and disaster risk reduction in Bhutan

3 Icy flood

Adapting to the impacts of glacial lake outburst floods in Bhutan

Introduction

In this chapter issues pertaining to disaster resilience in the mountainous country of Bhutan are examined, looking at a unique and deadly type of disaster – glacial lake outburst floods (GLOF) – understood to be linked to climate change-induced melting of glaciers. Much of the policy and practice in Bhutan relating to disaster resilience has its roots in the 1994 Luggye Tsho glacial lake outburst flood – the most massive GLOF disaster in recent history. The Himalayan-Hindu-Kush mountain range – one of the highest ranges in the world – runs through the northern part of Bhutan; its extensive glaciers have been retreating and transforming into large high-altitude glacial lakes due to global warming brought about by climate change. Thus, while the GLOF phenomenon appears unique to the Bhutanese topographical context, as climate change is a global phenomenon, the subject matter of this chapter has wider relevance.

In many ways, Bhutan is a microcosm of the transformations happening in other countries of the region. Its institutional and community efforts to achieve disaster resilience resonate with similar relatively recent efforts in the region, often spurred by large disasters (see for example Ebrahim, 2019, for GLOF disaster preparedness in Pakistan). Yet there are unique lessons to be learned from Bhutan, especially where the government has taken a lead role and forged effective partnerships with international bodies to address the critical challenge of implementing an integrated approach to climate change adaptation and disaster risk reduction. This is presented in this chapter through a case on the 1994 GLOF in Bhutan.

This chapter consists of two main parts: in the first part the context is presented by focusing on aspects relevant to disaster resilience including critical issues in Bhutan; the disasters affecting the country; and the country's administrative and institutional structure. The second part presents a case study of one of the most significant initiatives to address the GLOF disaster risk – an extensive, multi-year and multi-pronged project implemented in partnership with prominent international partners. This initiative offers a set of lessons for future disaster resilience policy and practice in the nation; much of the institutional and policy arrangements discussed in the earlier part stem from this project. Despite the challenges and limitations outlined, the project offers valuable broader lessons, particularly for other countries in the Himalayan-Hindu-Kush region that are also confronted with the spectre of GLOF disasters.

As in the other chapters in this book, the research methodology typically consisted of two main elements: a literature review of a variety of relevant publications, including official grey documents and even newspaper articles that reported on scientific information; and fieldwork in Bhutan to collect data. For limitations of transport, accessibility and time, it was not possible to visit the remote case study site in the Lunana region. It was possible to visit two of the areas affected by the 1994 GLOF, Punakha and Wandue, where early warning systems (EWS) had been installed in the case study project. Data collection techniques included stakeholder and key informant interviews, community meetings, transect walks, observation, and photographic and audio-visual recording. The field data and literature were analysed through a content analysis and then grouping of themes, coding and cross-referencing, which forms the basis of the discussion below.

Disaster overview: 1994 glacial lake outburst flood (GLOF)

The Lunana region located in the Pho Chu river sub-basin in north-western Bhutan includes many of the glacial lakes in the country including some of the ones posing the greatest disaster risk. As glaciers continue to melt and retreat, many of the closely spaced lakes expand and merge, forming large water bodies with significant volumes of water, hence developing potential for devastating glacial lake outburst floods (GLOFs) (Mool et al., 2001; Singh, 2009). Luggye Tsho is one such lake. In 1958 it was not visible on the topographical map and only became visible as a cluster of small lakes in 1967, which subsequently expanded and joined to form a single lake. By the early 1990s, the lake had doubled in size from a decade earlier, stretching to more than 2 kilometres in length (Ageta and Iwata, 1999). Luggye Tsho burst in October 1994, leading to the most catastrophic GLOF in Bhutan in recent history. The event resulted from a partial breach of the lake's impounding moraine due to the pressure from a gradual increase in volume and level of water from melting ice (Watanabe and Rothacher, 1996); accumulation of snow from the preceding winter may also have contributed. The water then flowed down into the adjacent lake, Raphstreng Tsho, and damaged its moraine wall; this released almost 18 million cubic metres of water which quickly gushed downstream (Mool et al., 2001; Singh, 2009).

Watanabe and Rothacher (1996) provided an account of the disaster: the GLOF water poured into the adjacent Pho Chu river and travelled downstream and caused severe devastation. The initial impact was on the first villages, located about 4,000 metres above sea level, with heavy damage to housing and infrastructure; the proximity of these settlements hardly allowed any time for warning and evacuation (USD Enterprises and Sutron Corporation, 2010). The destruction of a bridge marooned the villages, making it very difficult to provide emergency support. The major impact was in the more highly populated Punakha-Wangdue valley, almost 90 kilometres further downstream and almost 200 kilometres from the source; descending nearly 3,000 metres, the flood gained great momentum (Gurung et al., 2017) and caused extensive damage, especially in Punakha town where the Pho Chu river meets the Mo Chu river. After causing great destruction around this point of confluence of the two rivers, the flood then dissipated gradually with no further impacts downstream.

It took about seven hours for the GLOF water to travel downstream during the night to Punakha after it was first experienced in the upper settlements, hence there was an opportunity to provide early warning. There was a wireless radio communication system in place (Ives et al., 2010; Singh, 2009), but as it was late in the evening, staff were not available to receive the warning being transmitted from above, where the GLOF was already wreaking havoc. Thus, the massive gushing torrents carrying large amounts of debris, felled trees and logs, and descended on Punakha in the early morning and caught people unawares (United Nations Development Programme, 2016; Watanabe and Rothacher, 1996). A local resident narrated his experience:

> The flood happened without warning and we were awakened by loud noises. Our house was located on a nearby hill and we could see below that waves of more than two metres carrying many logs and other things hitting the houses on the riverbank. The water spread more than 15 metres on the two sides of the river. Many elderly people were very badly affected because they couldn't climb easily and move to safety.

The deluge killed 22 people and injured many others; the body of one person was washed downstream as far as Bangladesh (Watanabe and Rothacher, 1996). According to a survey conducted by the Department of Geology and Mines in the Punakha-Wangdue valley (Geological Survey of Bhutan, 1994), impacts were wide-ranging: 12 houses and five water mills were damaged; nearly 1800 acres of agricultural and pasture land were affected; livestock including 16 yaks were killed; 36 cowsheds with utilisable manure were lost; 6 tonnes of food grains were destroyed; four bridges collapsed; several religious buildings were destroyed or damaged; and the important Punakha Dzong (religious and administrative building) located at the confluence of the Pho Chu and Mo Chu rivers was seriously damaged (see Figure 3.1). The full extent of the devastation is not revealed by these figures from the downstream impact; the upper regions also incurred severe impacts. The economic loss amounted to millions of US dollars (Gurung et al., 2017).

There were also other impacts with long-term implications: the muddy and debris-filled water killed a vast amount of fish and it took more than a decade for the fish stock to return to normal (Watanabe and Rothacher, 1996). The impacts of the loss of livestock and arable land was also long term. Stones and silt deposits from the flooding rendered the flooded land unusable for future agriculture, requiring compensation from the government, although the newly allocated lands were not always suitable or in apposite locations (Wangdi and Kusters, 2012). The national economy is greatly dependent on agriculture including both farming and fishing, contributing to an estimated 17 per cent of GDP (Business Bhutan, 2011), much of it in the form of subsistence agriculture.

Figure 3.1 Punakha Dzong at the confluence of Pho Chu and Mo Chu rivers was damaged by the 1994 GLOF

Thus, the impacts at the community and low-income household levels, as revealed in a study by Wangdi and Kusters (2012), were highly detrimental to their long-term well-being and necessitated often desperate coping measures. By demonstrating the extreme risk posed by Bhutan's glacial lakes, the 1994 disaster led to further and more extensive long-term institutional initiatives for building resilience against GLOFs. This is why it has been discussed here to provide a background to the case study of a key resilience initiative presented later in this chapter.

Country overview of key issues

Bhutan is a mountainous land-locked country bounded by China to its north and India in the south (see Figure 3.2). The Himalayas mountain range passes through its northern region where elevations exceed 7,000 metres above sea level, dropping to about 150 metres in the southern foothills (Mool et al., 2001). It has the smallest area among the South Asian countries and the smallest population after the Maldives (World Population Review, 2019a). Some of the key issues that are critical to developing resilience in Bhutan are discussed below.

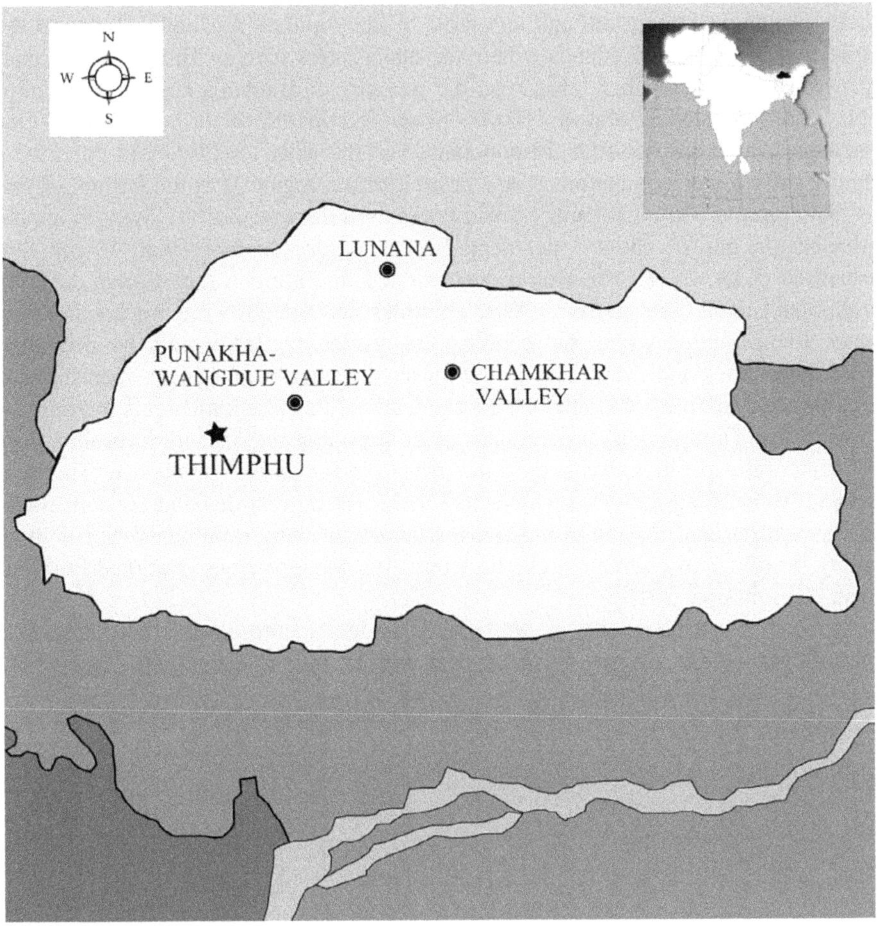

Figure 3.2 Map of Bhutan showing locations relevant to this chapter

Urbanisation

Bhutan has a population of about 800,000, which is low compared to most other South Asian countries. Its population density is among the lowest in the world with 20 persons per square kilometre (Civil Intelligence Agency, 2018; Preventionweb, 2014; World Population Review, 2019a). Nonetheless, the population has been growing steadily as indicated by the national population censuses. In the 2005 census the population was indicated as 634,982, which grew to 727,145 in the 2017 census, representing a 15 per cent growth in 12 years (National Statistics Bureau of Bhutan, 2006, 2018); a growing proportion of the population lives in cities.

The rugged mountainous terrain, particularly in the northern Himalayan region, make human settlement difficult and most of the population is concentrated in the valleys of the central highlands, where the major cities such as Thimphu and Punakha have been established. However, the primacy of Bhutan's capital city, Thimphu, is pronounced at almost 100,000 people compared to the next largest city, Punakha, which has one-fifth the population; all the other top 20 largest cities have significantly lower populations. Gasa in the Lunana region is at the bottom of this list with around 600 inhabitants (World Population Review, 2019b). With an annual urbanisation rate of about 3 per cent, Bhutan is already more than 40 per cent urbanised (CIA, 2018), most of it concentrated in Thimphu (see Figure 3.3). As with most capital cities in South Asia, Thimphu has been growing much faster than other urban centres within the country, the growth fuelled largely by migration from other cities and rural areas in the quest for better employment opportunities, facilities and services. The obvious consequence of this phenomenon is pressure on resources in Thimphu and a depopulation of the wider region with a resultant loss of skills and capacities, weakening the prospects for disaster resilience. This was evident from the agriculture sector where there was a great demand for conversion of agricultural land on peri-urban areas, whereas in many rural areas land is lying empty from lack of cultivation due to people migrating to cities (National Environment Commission, 2016).

Much of the urbanisation is unplanned, evident in the extensive construction of buildings close to rivers where they are at risk (Barrett and Lhamtshok,

Figure 3.3 Thimphu, the capital of Bhutan, is a rapidly growing city

2016; Meenawat and Sovacool, 2011) and on hilly slopes exposed to landslides and other hazards. There is a demarcated 'red zone' along riverbanks prohibiting construction (Kuensel, 2012b), but this is often disregarded due to the demands of urbanisation (Barrett and Lhamtshok, 2016). This is evident even in cities such as Punakha, which was severely impacted by the 1994 GLOF disaster. Even important public facilities such as hospitals and schools are being built in such locations; notably a US$760 million hydroelectric power plant and even an airstrip are located in flood-prone areas (Meenawat and Sovacool, 2011; Nayar, 2009). Such developments frequently tend to experience small-scale hazard events which do not attract much institutional attention (Komori et al., 2012). Bhutan does have building regulations and codes that have recently been updated (Department of Human Settlement, 2017), but it was reported by interview respondents during fieldwork that these are enforced mainly in central urban areas and often disregarded in the smaller towns and villages. As noted by Barrett and Lhamtshok (2016, p. 21), 'There is also insufficient and slack enforcement of building by-laws codes in the urban areas.' Thus, even a small country with currently a low population is beginning to experience urbanisation pressures similar to other countries in the South Asian region.

Transport and accessibility

There is no rail or boat transport in Bhutan – transportation by road is the main option. There is a road network connecting the main urban centres (Trigona, 2018), but the mountainous topography acts as a constraint to wider accessibility. Settlements in the mountains tend to be sparse and scattered and many are remote, particularly in the northern Himalayan region. Access to that region involves pedestrian movement or animal transport and is highly inaccessible. There are ongoing undertakings to improve the road network (Kuensel, 2016a, 2016b) but road transport can be slow, difficult and hazardous because of the hilly terrain (Himalayas, n.d.), especially in areas away from the main cities. Road conditions are often poor (Dema, 2016) and most outer roads are narrow and single-lane (Trigona, 2018). Such conditions pose serious socio-economic developmental obstacles for the remote, rural areas, and providing emergency and disaster recovery services can be challenging (Barrett and Lhamtshok, 2016).

Environmental aspects

Bhutan's national policy stipulates that a minimum of 60 per cent of the country should remain under forest cover at all times; recent estimates indicate that the figure is more than 80 per cent (NEC, 2016). It is widely mentioned that Bhutan is the only country in the world with negative carbon emissions because of the great absorption capacity of its forests (Climate Council, 2017; Mellino, 2016; Tutton and Scott, 2018). There are a wide range of environmental initiatives in Bhutan, including generating all of its own domestic electricity and even exporting power to India; initiatives include hydroelectric plants, energy-efficient vehicles and conservation of forests and biodiversity (Nayar, 2009; Pashley, 2015). Bhutan appears to be at the forefront of environmental sensitivity and action.

However, as the most recent *State of the Environment Report* of Bhutan's National Environment Commission (NEC, 2016) pointed out, population growth and urbanisation are beginning to place significant pressures on the environment. These transformations are accompanied by economic growth, with Bhutan among the fastest growing economies in the world with more than 7 per cent annual GDP growth (CIA, 2018). This growth results in higher consumption patterns and greater demands on the environment, which in turn leads to greater waste production – highly visible from the widespread littering and mass dumping at settlement peripheries. Solid waste is beginning to pose a serious environmental problem given the limited capacity and awareness for its management (Gyelmo, 2018). A range of other emerging environmental problems were brought to attention by the NEC (2016) report: demand for land for urban development, including building roads and farming, which deplete the forests; commercial timber logging; extraction of fuel wood, reportedly at one of the highest consumption rates in the world due to the vast majority of the rural population being reliant on it; pollution of natural water bodies, especially rivers, by direct discharge of sewage and effluents; air pollution in the cities; and illegal poaching endangering biodiversity.

Thus, despite being a champion for the environmental cause, Bhutan is beginning to face a similar set of challenges from the development versus environment dilemma confronting all the South Asian countries; indeed, this is a global quandary pervasive throughout the developing world. There is a link between such unsustainable 'development'-driven environmental degradation and disaster risk. For example, activities such as construction of roads and mining involve excavation and cutting down hills (see Figure 3.4), leading to landslides and flash flooding (Barrett and Lhamtshok, 2016); urban densification prompting construction of buildings near rivers creates exposure to floods, as mentioned above when discussing urbanisation.

Disaster risk profile

Compared to some of the nearby South Asian countries such as Bangladesh, India, Nepal and Pakistan, the hazards in Bhutan have proportionately less impact (Lotay, 2015). This is possibly so because of the lower population and the characteristically small dispersed settlements in the hilly terrain. Despite the rapid urbanisation discussed above in the overview of key issues in the country, the population of the largest city, Thimphu, is about a tenth of that of Kathmandu in Nepal (World Population Review, 2019c), and much less than a hundredth times of the megacities of the region such as Dhaka, Mumbai or Karachi, which are confronted with high disaster risks. Hence, even in the more densely populated urban areas of Bhutan, disaster impacts are less pronounced.

Nonetheless, Bhutan is vulnerable to various hazards because of its mountainous geography, which consists of steep slopes necessitating the location of settlement, agriculture and transport networks primarily in the river valleys;

Figure 3.4 Cutting down hills, such as this one in Thimphu, creates the risk of landslides

seismically active geological conditions; great elevational difference within different regions; regional and seasonal temperature differences ranging from tropical to sub-zero; and unpredictable and changing climatic conditions. Two of the key economic sectors – hydroelectric power generation and agriculture (Bouissou, 2011; Business Bhutan, 2011) – are highly sensitive to hazards (Meenawat and Sovacool, 2011; Tshering, 2017). In a way, it is fortunate that the large Basochhu Hydropower Plant downstream of Punakha was built after the 1994 GLOF (Druk Green Power Corporation, 2019) and hence not impacted, but future GLOFs pose a serious risk to this important facility and other hydroelectric plants in the country.

This disaster profile of Bhutan has been prepared by consulting some key references (Barrett and Lhamtshok, 2016; Lotay, 2015; Meenawat and Sovacool, 2011; Norbu, 2012; Pradhan, 2014; Preventionweb, 2014; Wangchuk, 2012) and also from information collected from different key informants during fieldwork in Bhutan in 2018. Other supplementary sources have been cited below as relevant.

Earthquakes

Bhutan is located in a seismically active zone. The eastern part is most at risk, along with parts in the west that are also at high risk; different levels of earthquake risk prevail throughout the country (Tshering, 2017). Estimates for mortality and economic losses due to earthquakes for the period from 1990 to 2014 amounted to 46.9 per cent and 82 per cent respectively, exceeding the impacts of all the other hazards (Preventionweb, 2014). There is a history of earthquakes causing extensive casualties and damage; recent ones included four in the 2000s. The 2009 and 2011 earthquakes – both exceeding magnitude 6 on the Richter Scale and the 2011 reaching almost magnitude 7 – resulted in heavy impacts with economic loss exceeding US$76 million according to reports jointly produced by the Royal Government of Bhutan, World Bank and the United Nations (2009, 2011). Key national sectors were heavily affected in these two earthquakes, 13 persons were killed and many more were injured.

The above reports indicate that the housing sector was the most severely affected, as globally evident in many disasters (see for example Marti, 2005); this was also mentioned in the chapter on the Maldives in this book. More than 12,500 houses were affected, ranging from partial damage to complete destruction. Unreinforced brick, stone and earth structures are widespread; such constructions are highly vulnerable to earthquakes, hence contributing to a heavy toll. However, provisions for earthquake resilience is largely absent in the national building codes (for example, see the presentation by Department of Human Settlements, 2017). It was only recently that capacity building initiatives for government staff in this field have been initiated by an external organisation (Build Change, 2016), and some training programs for masons and carpenters have reportedly begun.

An important issue is the link between earthquakes and other hazards. As discussed below, landslides constitute a serious hazard in this mountainous context and earthquakes can trigger landslides as demonstrated in recent earthquake events. Additionally, earthquakes have the potential to activate GLOFs (Bellamy and Penjor, 2014; Nayar, 2009). This points to the need for a multi-hazard approach to developing disaster resilience, as well as an understanding of the interacting nature of different hazards and how a primary hazard can have cascading secondary impacts.

Landslides

The mountainous topography of Bhutan is susceptible to frequent and extensive landslides. Although the hazard persists throughout the country, the eastern and southern regions are more vulnerable because of the steep slopes with small, broken rocks. There are regular reports on landslides in local daily newspapers, especially during the monsoon rainy season (see for example Dema, 2018; Rai, 2018). In the process of urbanising, mountains are cut to make way for roads and settlements, exposing the slopes and making them prone to landslides; this was pointed out by several key informants during the fieldwork in Bhutan.

Heavy rainfall is one of the key factors that precipitate landslides and most landslides occur annually during the rainy season between May and September. Landslides consist of movement of soil and rockfalls (Penjore and Sato, 2012), the latter posing a great risk to human lives, vehicles and property. Road blockages and traffic disruption by landslides is common, which in turn have economic impacts by affecting the supply chain. In fact, damage to roads and consequent repair and maintenance constitutes one of the key government expenditures (Pradhan, 2014). Agriculture is another key sector affected by landslides as they cover fields with rocks and soil, making them unusable for a long time afterwards. Interestingly, it has also been pointed out that historically landslides have created terraced land for settlement and agriculture and in that sense the Bhutanese people had benefited (Penjore and Sato, 2012). However, in the present age, such benefits appear to be far outweighed by the above-mentioned impacts by landslides on different national sectors.

The interacting nature of different hazards is a key aspect of landslides. Earthquakes and seismic tremors can trigger landslides – for example, in 2011 an earthquake in the neighbouring state of Sikkim, India, caused massive landslides in Bhutan, killing one person (Lotay, 2015). Forest fires also denude hillsides of vegetation and expose the topsoil, leading to landslides (Wangchuk, 2012), and the soil then flows into rivers and raises riverbeds, making the rivers more flood-prone. This also hampers hydroelectricity production (Gyelmo, 2016) which is essential to the national economy. Importantly, landslides and flash floods are often concurrent as both can be triggered by heavy rainfall. This is also discussed in the chapter on Sri Lanka in this book.

Fortunately, because of the typical occurrence during the rainy season, there are opportunities for preparedness and evacuation, allowing for the loss of human lives to be avoided. Compared to earthquakes and floods that resulted in 46.9 per cent and 37.5 per cent mortality respectively during the period from 1990 to 2014, landslides only accounted for about three per cent mortality (Preventionweb, 2014). Nonetheless, the National Adaptation Programme of Action (NAPA) II included landslide risk mitigation as a key area of action because of the heavy impacts on the economy and other development sectors (ChhimidD Consulting, 2018; Lhamo, 2018; Royal Government of Bhutan and United Nations Development Programme, 2014). Moreover, landslides have become a serious hazard due to the intrinsic link between climate change and changing rainfall patterns, leading to heavier, more intense, unpredictable and erratic monsoon rains (Meenawat and Sovacool, 2011; United Nations Environment Programme, 2009). This is why it is taken into consideration in the NAPA II as a 'climate-induced disaster' with a view towards the future when the hazard can be expected to increase. A key area of research suggested during the fieldwork in Bhutan by scientists in the National Centre for Hydrology and Meteorology is to understand the link between weather conditions and landslides – particularly the rainfall thresholds that trigger landslides – in order to develop an effective early warning system.

Windstorms

Strong windstorms accompanied by lightning, hail, heavy rain or snow in higher altitudes are common in Bhutan, especially during the spring and autumn seasonal changes. This hazard seems to be increasing because of climate change, resulting in more frequent and intense events. Such windstorms cause great damage; although human casualties were much less than other disasters in the period from 1990 to 2014, the economic losses were high at 15.2 per cent (Preventionweb, 2014). The windstorms of 2011 and 2013 had the heaviest impact, which together damaged almost 3,500 houses and many other public and religious buildings.

A key impact of windstorms is on buildings; from 2010 to 2017, 5,500 houses and other buildings were affected by windstorms (Department of Engineering Services, 2017). The maximum impact is on the roof, which tends to blow off (Lotay, 2015). It is common to see roofs of buildings in Bhutan weighed down by stones to prevent displacement by wind – a local coping mechanism with limited efficacy. For this reason, under the NAPA II project mentioned above in the sub-section on landslides, a key output was the development of a training manual on wind-resistant roofing (ChhimidD Consulting, 2018) produced by the Department of Engineering Services (2017). However, wind-resistant construction is not included in the national building codes (Department of Human Settlements, 2017) and there does not seem to be a strategy for widespread dissemination and application of the manual's recommendations, hence the vulnerability of buildings to windstorms can be expected to prevail.

It should be noted that although Bhutan is a land-locked country, it is vulnerable to strong cyclones from the Bay of Bengal more than 1,500 kilometres to the south, as evident from the impact of Cyclone Aila in 2009. While the cyclone caused maximum devastation with its ferocious wind and surges in the coastal areas of Bangladesh and West Bengal, India, (International Life Saving Federation, 2009), in Bhutan the impact was experienced in the form of heavy rainfall in almost all 20 districts. This triggered massive floods and landslides; in Punaka and Wangdue, areas affected by the 1994 GLOF, the flood level was on par with that of the GLOF event (UNDP, 2012a). At least four persons were reported killed (Lamsang, 2009), more than 4,600 households were affected and losses amounted to more than US$17 million (UNDP, n.d.; World Bank, n.d.). Even the GLOF risk mitigation work discussed below in the case study was disrupted by the cyclone's impact (Business Bhutan, 2009). This event demonstrates the transboundary nature of disasters and therefore the requirement for disaster resilience initiatives to have a systems approach considering the wider regional risk catchment beyond national borders.

Forest fires

While being extensively forested is an environmental benefit for Bhutan (see sub-section on environmental aspects above), it also creates the risk of forest fires. On average, 10,000 acres of forest are burnt annually (Barrett and Lhamtshok, 2016); in 2014, it was reported that more than 20,000 acres were lost (Pradhan, 2014). Between 2008 and 2014 there were more than 300 fire incidents burning nearly 93,000 acres of forests (Lotay, 2015). Forest fires cause serious damage to biodiversity and the local ecosystem, infrastructure and property. Hydropower stations are also at risk (Gyelmo, 2016) and damage to them could cripple the economy. The shortage of water in the upper altitudes present a constraint in fire-fighting, and because of their remote locations it is difficult to provide emergency services.

Forest fires usually happen during the dry season from November to April; because of the propensity for landslides and floods to occur during the rainy season, organisations and communities in Bhutan have to deal with some hazard or the other throughout the year. As mentioned above in the sub-section on landslides, by burning off vegetation that binds the soil, forest fires make the mountain slopes prone to landslides. However, as landslides occur during the rainy season, it allows a few months' time to prepare after the forest fires have occurred. All the forest fires are reported to be ignited by human activities such as agriculture and cooking, including both accidents and intentional lighting (Gyelmo, 2016). There is thus scope for building awareness in communities that live near forests and rely on forest-based livelihoods.

Human-induced hazards

The mountainous geographical setting of Bhutan renders road conditions difficult, as discussed above in the subsection on transport and accessibility in the section of key issues in the country. This leads to a large amount of road traffic accidents, attributed to reckless and drunken driving, poor road and weather conditions, and vehicular malfunctions (Pradhan, 2014). As estimated by Wangdi et al. (2018), there were more than 1800 accidents resulting in more than 150 deaths and over 11,000 injuries in a single year between 2013 and 2014. In recent times there have been governmental initiatives to improve road conditions and better vigilance, reportedly resulting in a 40 per cent reduction in accidents from 2011 to 2018 (Phuntsho, 2018), which allows a positive outlook for a safer future on Bhutan's roads.

In addition to forest fires, there are many fire incidents in settlements due to negligent cooking with wood, faulty electrical systems and other forms of human carelessness. Of note is the destruction in 2012 of the centuries-old historical fortress, Wandguephodrang Dzong, resulting in the great loss of a cultural heritage building (Lotay, 2015) and many valuable artefacts and documents inside.

Epidemics and pests also pose a risk, but although there have been some pestilence cases, such as Army Worm infestation in 2012 (Pradhan, 2014) and small outbreaks of Avian Influenza, these seem to be managed and controlled effectively by the government (Lotay, 2015; Reuters, 2018; The Poultry Site, 2016). Typical developmental problems, such as waste mismanagement and environmental degradation, as discussed above in the section on key issues in the country, interact with natural hazards to create complex scenarios that inhibit resilience in Bhutan.

Climate change impacts

As Bhutan is a land-locked country, the challenges faced by coastal regions due to climate change – a field of strong institutional and academic interest and engagement – do not apply directly here. Nonetheless, increasingly ferocious cyclonic storms resulting from climate change can cause impacts from coastlines thousands of kilometres away, as was the case with Cyclone Aila in 2009, discussed above in the sub-section on windstorms. Climate change is thus presenting unprecedented hazards in Bhutan, requiring newer approaches towards supporting resilience.

A key climate change-related issue in Bhutan is the thawing of ice and permafrost in the high-altitude Himalayan region, and the retreat and diminution of glaciers, leading to the risk of GLOFs, as discussed below in the sub-section on floods. As succinctly summarised by Dorji (2017, p. 58), 'retreating glaciers and the formation of supraglacial lakes are among the most significant climate change impacts observed in recent years.' Different sources report that large blocks of melting ice are breaking off from glaciers on a regular basis and landing in glacial lakes, contributing to the potential for GLOFs (for example see Goldenberg, 2011; Nayar, 2009; Singh, 2009). Various scientific studies warn that with current global warming projections, more than half of the glaciers in Bhutan might be lost by the end of this century (UNDP, 2012a). Despite the widely observed impacts of climate change on the Himalayan glaciers, it was reported in a study based on historical long-term trends that GLOFs are not increasing in frequency (Komori et al., 2012). However, whether this would remain true in the face of advancing climate change can be questioned.

Reduction of the risks of GLOF disasters is thus understood as a climate change adaptation response, evident from the case study discussed below implemented under the National Adaptation Programme of Action (NAPA) I. Indeed, it was the first project funded by the internationally prominent Global Environment Facility for climate change adaptation (UNDP, 2012a). Thus, participants at an important international conference in Bhutan on GLOFS articulated the need for a combined approach to climate change adaptation and disaster risk reduction (UNDP, 2012a).

A wide range of impacts of climate change on different sectors in Bhutan can be anticipated. The continuing loss of glaciers can contribute to instability of water supply (Fujita et al., 2012). This impact of the warming climate is already being experienced in the form of drought and shortage of water, especially at the upper altitudes; this has strongly affected agriculture and forestry and has also created conditions for forest fires to spark easily (Lotay, 2015; Meenawat and Sovacool, 2011), another example of interacting hazards. Similarly, global warming might result in more avalanches and landslides (United Nations Environment Programme, 2009). The reduction in water availability can also be anticipated to hinder the production of hydroelectricity (UNDP, 2012a) and thereby affect a key economic sector in Bhutan; this would also impact an environmentally sustainable source of energy production that contributes towards climate change mitigation.

The instability of water regimes in the Himalayas might have far-reaching downstream repercussions in Bangladesh and India, where the basins of the rivers originating in the Himalayas are located – more than a billion people rely on water from these sources. Thus, global warming is an important transboundary issue in the region, extending beyond the borders of Bhutan.

Floods: GLOFs and rain-fed

The retreat of glaciers, accelerated in recent times by global warming, has resulted extensively in the formation of high-altitude supraglacial lakes at 4,000–5,000 metres above sea level in mountain ranges across the world such as the Hindu Kush and Himalayan ranges (Awale, 2019; Ives et al., 2010; Mool et al., 2001; Singh, 2009; UNEP, 2009; United Nations Development Programme, 2012a). It is estimated that glaciers in the Himalayas are retreating at the rate of 30–60 metres per decade, resulting in expanding volumes of water in the glacial lakes (Bellamy and Penjor, 2014; Gerretsen, 2019; Morelle, 2019; United Nations Development Programme, 2007). Typically, these lakes are impounded by rock walls or formations called moraines which usually have an ice core. Melting of the ice can result in release of the lake's water, causing a GLOF and downstream flooding, as discussed above in the section on an overview of the 1994 GLOF disaster.

Other factors that can trigger a GLOF by causing a breach of a glacial lake's moraine include earthquakes, avalanches, rockfalls, rupture of a glacier (calving) and heavy snowfall. Extreme cases involve rapid release of millions of cubic metres of water, causing heavy damage and casualties (Gurung et al., 2017; Kattelmann, 2003). In recent history in the Hindu-Kush-Himalayan ranges, 56 GLOF events were recorded (UNDP, 2012a); Bhutan is one of the countries in the region that is highly vulnerable to such events. There are 2,674 glacial lakes in Bhutan, of which 22–25 pose a risk of GLOFs (Gurung et al., 2017; Ives et al., 2010). The case study presented below is about one of the most significant GLOF risk reduction projects to be implemented in Bhutan, and indeed, in the wider Hindu-Kush-Himalayan region.

The other form of floods that affect Bhutan are due to heavy rainfall during the annual summer monsoon season during May to September. Because of the hilly topography these floods can rapidly gain momentum and become flash floods, that, along with landslides, cause great damage, as discussed above in the sub-section on landslides. The hilly terrain necessitates settlement and agriculture to be carried out near rivers because of the difficulty of water supply uphill, hence the exposure to both GLOFs and rain-fed floods.

During the fieldwork in Bhutan, scientists at the National Centre for Hydrology and Meteorology explained some key differences between GLOFs and rain-fed floods. At the early stage, GLOFs can cause massive devastation, but as the waters flow downstream they become attenuated and lose their destructive power. On the other hand, rain-fed floods gain power with increasing rainfall and cause massive destruction as they travel downstream; the southern and eastern parts of the country with lower elevations are more affected. The southern city of Phuntsholing thus experiences great impacts of such floods and associated landslides, which is why the NAPA II programme mentioned above in the sub-section on landslides focused its floods and landslides mitigation work in that area (ChhimidD Consulting, 2018;

Lhamo, 2018; RGoB and UNDP, 2014). The other difference is that it is difficult to predict when a GLOF might happen – it might suddenly impact on one fine morning as the 1994 GLOF did in Punaka (see overview of the disaster above). Rain-fed floods, by comparison, are seasonal, often slow-onset and can be predicted when there is incessant and heavy rainfall during the rainy season; thus, early warning, evacuation and preparedness plans can be implemented. Additionally, it is commonly believed that GLOFs occur in the monsoon season because of heavy rainfall, but even in the dry winter season, a GLOF might possibly happen because of an earthquake (Bellamy and Penjor, 2014). There is therefore a need for continuous year-round vigilance and monitoring for GLOFs.

Policy aspects of disaster resilience

In line with disaster resilience trends both globally and in other South Asian countries, the policy framework for disaster resilience in Bhutan was propelled by major disasters impacting the country, including the 1994 GLOF and the 2009 and 2011 earthquakes (discussed above in the section on disaster profile of Bhutan). Nonetheless, climate change is considered a serious issue in Bhutan; some of the key programs for disaster resilience are conceptualised in terms of adapting to the negative impacts of climate change under a series of NAPAs beginning from 2005. The disaster resilience administrative structure and institutional arrangements in Bhutan have been described by several sources (Barrett and Lhamtshok, 2016; Lotay, 2015; Norbu, 2012; Pradhan, 2014; Wangchuk, 2012), which have been drawn from here. The main aspects of the policy framework and its institutional structure are discussed below.

Administrative structure

The government administrative structure of Bhutan operates at the national and local levels and most national government agencies are based in the capital, Thimphu. The country is divided into 20 *Dzongkhags* (equivalent to districts) and the local government administrative sub-levels consist of *Thromdes* (municipalities), *Dungkhags* (sub-districts) and *Gewogs* (blocks). Only the larger *Dzongkhags* include *Dungkhags* and the *Gewogs* are further sub-divided into *Chiwogs* (sub-blocks) that in most cases comprise several villages. At each level or sub-level, the governing body is the *Tshogde* composed of elected members and headed by a *Thrompon* (administrative leader). The Local Government Act 2009 (Parliament of Bhutan, 2009) is an initiative towards further decentralisation by updating previous acts and providing legal governing directives for all the above administrative levels.

Institutional arrangement for disaster resilience

An organogram of Bhutan's institutional structure for disaster resilience is shown in Figure 3.5. The National Disaster Management Authority (NDMA), which

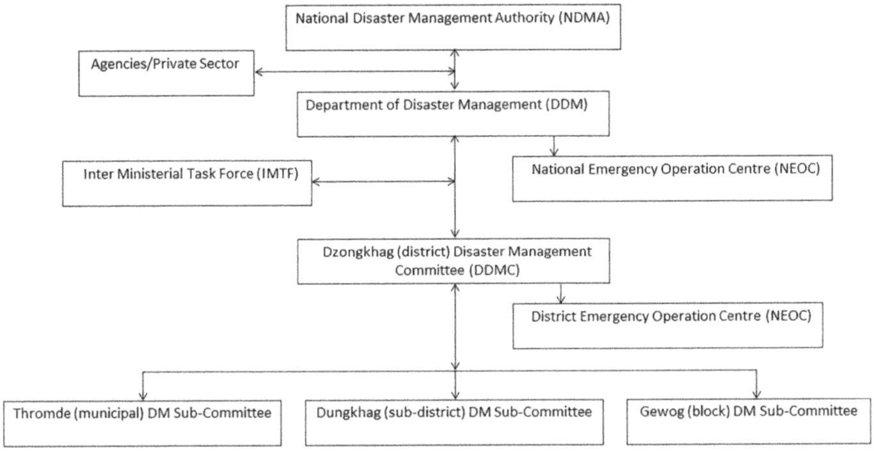

Figure 3.5 Disaster resilience institutional structure in Bhutan (adapted from: Lotay, 2015; Norbu, 2012)

came into force legally through the Disaster Management Act 2013 (National Assembly of Bhutan, 2013), is the highest decision-making body for all activities relating to disasters in the country. Under the leadership of the Prime Minister, the members of NDMA are drawn from relevant ministries, and other public and private sector agencies. In addition to providing directives, approvals and oversight to its subsidiary bodies, NDMA is responsible for budget allocation for disaster resilience activities. It also engages in capacity building initiatives.

The Department of Disaster Management (DDM) under the Ministry of Home and Cultural Affairs (MoHCA) is the Secretariat and executive arm of NDMA and is responsible for coordinating and translating policies and directives to the different administrative levels.

DDM has three main divisions corresponding to its main areas of action: preparedness and mitigation (pre-disaster); response and early warning (during disaster); and recovery and reconstruction (post-disaster). DDM is advised by an Inter-Ministerial Task Force (IMTF) composed of experts from different ministries, which provides support for preparing and implementing the *National Disaster Management and Contingency Plan*. The national policies place emphasis on mainstreaming disaster resilience into all national sectors, and the IMTF is an important body for that purpose.

At the local level, in each *Dzongkhag* there is a Disaster Management Committee, and in each *Thromde*, *Dungkhag* and *Gewog* there is a Disaster Management Sub-Committee. The sub-committees are guided by the *Dzongkhag* Disaster Management Committee (DDMC), which in turn is guided by DDM. Among a range of responsibilities, DDMC prepares and implements the *Dzongkhag Disaster Management Plan* and *Contingency Plan* in alignment with the national level plans, and also supports the sub-committees at the lower levels to produce and run their disaster management and contingency plans. Under the Critical Disaster Management Facility, there is a National Emergency Operation Centre (NEOC) coordinated by DDM and under the NEOC there is a *Dzongkhag* Emergency Operation Centre (DEOC), and in some cases there are also emergency operation centres at the lower administrative level. These centres primarily deal with actions during emergencies, including early warning, medical assistance and search-and-rescue.

Policy instruments and challenges

The following main policy instruments relate to disaster resilience in Bhutan:

- *National Disaster Risk Management Framework* (NDRMF), established in 2006. The 1994 GLOF disaster was a significant event that pointed to the need for a policy instrument to address disaster resilience. However, Bhutan had limited experience in this field and it thus required more than a decade to prepare and adopt the NDRMF.
- *Disaster Management Act 2013*, which was driven by the massive earthquakes of 2009 and 2011 (see the section on disaster profile above) and also the GLOF risk reduction project discussed below as a case study. This policy instrument with a legal basis repealed the previous NDRMF. It is the most important policy instrument in Bhutan.
- *National Disaster Risk Management Strategy* (NDRMS), a strategic derivative of the DM Act 2013. It is aligned with the national five-year planning process and the global *Sendai Framework for Disaster Risk Reduction*.
- *Disaster Management Rules and Regulations 2014*, which are derived from the DM Act 2013 provide a legal basis of all actions by agencies relating to disasters.
- *National Disaster Management and Contingency Plan*, which is prepared and implemented by DDM and translated into local level policy instruments by following the *Disaster Management Planning Guidelines 2014*. It was reported during fieldwork in Bhutan in 2018 that out of 20 *Dzongkhags* (districts), 16 had already prepared their disaster management and contingency plans.
- Two important national sectors – education and health – have specific policy instruments for disaster resilience in the form of the *National Action Plan for School Earthquake Safety* and *National Action Plan for Earthquake Safety of Health Facilities* to address their vulnerability to earthquakes.
- The projects under the series of *National Adaptation Programmes of Action (NAPAs)* were the most significant initiatives towards disaster resilience, discussed below, and reflected in the case study below in this chapter.

Despite the relatively recent start on disaster resilience planning in the mid-2000s, compared to some of the other South Asian countries, it is remarkable that such structured institutional arrangements and policy instruments have been developed in Bhutan. Yet there are many challenges, as reported by DDM staff during the fieldwork, and also in presentations by DDM personnel and other governmental staff (Norbu, 2012; Pradhan, 2014; Wangchuk, 2012). The main constraint in translating the elaborate and complex scope of the policy instruments into action is lack of capacity. Particularly at the local level DM committees and sub-committees, there is severe shortage of skilled staff and thus there is a reliance on top-down directives from the national level and very limited autonomous decision making and initiatives. Even at the national level there is limited technical capacity, especially for risk assessments; most such assessments are localised, but very little exists at a nationwide level. Coordination between the different levels as well as between the different sectoral stakeholders is also a challenge, which in turn creates barriers to mainstreaming disaster resilience across the different sectors. The lack of capacity is further constrained by lack of funds and other resources such as equipment and tools. It was reported that good quality and modern equipment for search-and-rescue operations in particular was greatly lacking. From 2010 to 2015 the annual budget allocation to DDM was on average US$0.75 million (Lotay, 2015) – a substantial amount in local terms, but even then, it does not seem adequate.

Bhutan is a highly traditional country and there are strong principles on maintaining cultural aspects in architecture, attire and etiquette, so it is somewhat unusual that cultural aspects are not referred to in the DM Act 2013 and other policy instruments. The culturally embedded fatalistic view that disasters are 'natural' and 'divine' acts is widely pervasive, especially in rural and semi- or non-literate communities, and there is thus a strong need to raise awareness of the potential for human agency to achieve disaster resilience; this should be the basis for national disaster resilience initiatives. The other important issue not acknowledged in policy is the rapid and often unplanned urbanisation taking place in Bhutan (see section above on key issues in the country), which requires a specific approach to disaster resilience beyond the somewhat generic and broad-based policy approaches that exist at present.

One of the main obstacles is the mountainous terrain, which makes it difficult to implement disaster resilience policies across the country – particularly at the local government level and in remote settlements in the Himalayas. Transport and supply of socio-economic services, including emergency services, presents serious challenges in this developing country context. Thus, despite the intensive institutional efforts to develop a structured approach towards enabling disaster resilience, it is not possible in the current context to realise it in its entirety.

National Adaptation Programmes of Action (NAPAs)

A recognition of the climate change impacts affecting Bhutan led to a series of National Adaptation Programmes of Action (NAPAs); GLOFs were considered one such critical impact. Thus, in the first NAPA initiated in 2005 (UNDP, 2007), resilience against GLOFs was specifically targeted in a project led by UNDP with a number of national partners. Being such a hallmark project that addressed one of the hazards posing possibly the greatest risk in Bhutan, it was selected as a case study for this chapter as presented in the next section.

Figure 3.6 Slope stabilisation work carried out under NAPA II (courtesy of UNDP-Bhutan)

The significance of the NAPA I project was reflected in the subsequent NAPAs. In the second, NAPA II (2014–2017), UNDP partnered with the National Environment Commission (NEC) on a project to focus on 'the risks of climate-induced disasters' (RGoB and UNDP, 2014, cover page), demonstrating a combined approach towards disaster risk reduction and climate change adaptation. While the main thrust was on building national and local capacity, and an extensive range of actions were undertaken towards that purpose, physical disaster risk mitigation work was also undertaken by stabilising hill slopes against landslides and building flood protection and erosion prevention structures along riverbanks (see Figure 3.6). Other hazards addressed in the project include forest fires and windstorms. An evaluation of the project indicated significant improvements in the different areas of action of the project (ChhimidD Consulting, 2018).

Since 2018, the Gross National Happiness Commission (GNHC) has partnered with UNDP to implement a six-year climate resilience project on sustainable agriculture, forestry and livelihoods (UNDP, 2017) under the third NAPA, NAPA III. Gross National Happiness (GNH) is the unique and fundamental governance philosophy of Bhutan and the GNH index is a key measurement tool for human development in the country (Oxford Poverty and Human Development Initiative, n.d.). The GNHC is the national government body for ensuring that all development policies and plans are aligned with the principles of GNH, including those for disaster resilience.

These NAPAs represent some of the most extensive initiatives in Bhutan to address the combined challenges of climate change and disaster risk. The future of disaster resilience in the country is inextricably linked to the outcomes of the NAPAs, which is why this chapter includes a case study from one of the NAPAs.

Case study: 'Reducing climate change-induced risks and vulnerabilities from glacial lake outburst floods in the Punakha-Wangdue and Chamkhar valleys'

Recognition of the critical disaster risk, particularly of GLOFs, posed by the high altitude glacial lakes in the Lunana region in north-western Bhutan due to the impact of climate change led to the project entitled 'Reducing climate change-induced risks and vulnerabilities from glacial lake outburst floods in the Punakha-Wangdue and Chamkhar valleys'. The project was initiated by UNDP under NAPA I, and ran from 2008 to 2012 in conjunction with the Department of Geology and Mines (DGM) and Department of Hydro-Met Services (DHMS) under the Ministry of Economic Affairs (MoEA), and the Department of Disaster Management (DDM) under the Ministry of Home and Cultural Affairs (MoHCA).

The bulk of the funding for the project was provided by the Global Environment Fund (GEF) through its Least Developed Countries Fund (LDCF), contributing nearly US$3.5 million. The Austrian Development Agency provided about US$800,000; UNDP about US$400,000; World Wildlife Fund Bhutan about US$30,000; and the Bhutanese government provided more than US$2.5 million in-kind contributions (UNDP, 2011a). This was one of the most significant projects of its kind in Bhutan, involving a large financial investment and a wide range of national and international stakeholders.

The project mainly implemented structural risk reduction measures, and also some non-structural measures including capacity building, early warning systems and knowledge dissemination. Key activities included: artificial lowering of the water level of Thorthormi Tsho, the largest and most high-risk glacial lake; institutional and community capacity building on disaster resilience; land-use planning informed by hazard zonation mapping; expansion of an early warning system along the Punakha-Wangdue valley; and different measures for disseminating the lessons gained from the project. The project demonstrates practical measures to reduce GLOF risks and provides long-term lessons.

Background to the project

The Lunana region in north-western Bhutan includes some of the most high-risk glacial lakes including Luggye Tsho, Rapstreng Tsho and Thorthormi Tsho (Singh, 2009). The 1994 Luggye Tsho GLOF, discussed earlier in this chapter when an overview of this disaster was presented, raised public awareness and instigated the government to undertake steps to increase resilience against such events; a team of experts from several countries was mobilised for that purpose. After preliminary studies, the team advised a set of risk reduction measures including structural interventions such as repairing the moraine wall of Raphstreng Tsho damaged by the outburst, building gabion-toe retaining walls along the Pho Chu river, and lowering the water level of Luggye Tsho and adjacent lakes by draining out water by canalisation, hence reducing the hydrostatic pressure. These actions were advised to be supplemented by long-term measures such as safeguarding against earthquakes, planting vegetation for slope stability and setting up meteorological and monitoring stations (Mool et al., 2001).

The *Raphstreng Tsho Outburst Flood Mitigation Project* was then initiated in 1996. Facing the immense difficulties of working in such a remote location largely with manual equipment, the ambitious scope of the project was toned down somewhat. Nevertheless, it was possible to lower the lake's water level by four metres in a couple of years (Singh, 2009) and bring the risk under a manageable threshold. Also, after losing a large volume of water in the 1994 GLOF, Luggye Tsho, became less hazardous (Kuensel, 2012a), though there is some residual risk because the lake still has a depth of 60–100 metres (Kuensel, 2011).

Phase two of the *Raphstreng Tsho Outburst Flood Mitigation Project* began in 1999 supported by Austro-Bhutanese cooperation. The main aim was to assess the geomorphological risks of Raphstreng Tsho and Thorthormi Tsho. The activities involved fieldwork with an integrated multidisciplinary approach using remote sensing, geological, hydro-geological and geophysical methods to interpret the subsurface characteristics of the moraine dam (Häusler et al., 2000; Mool et al., 2001). The investigations indicated that the risk of an outburst from Raphstreng Tsho was low but the risk from Thorthormi Tsho was high (Häusler et al., 2000).

Thorthomi Tsho had been expanding rapidly due to climate change-induced glacial melt; in 2001 it was 1.28 square kilometres, but in less than a decade it had grown to 3.42 square kilometres and become three times the size of Luggye Tsho (Kuensel, 2010; Singh, 2009). Raphstreng Tsho by itself posed little risk after the structural mitigation work done on it; however, as it is about 80 metres below Thorthormi Tsho and separated by an unstable ice-core moraine of about 32 metres at its thinnest part, it could overtop significantly if an outburst occurred from Thorthomi (Bhutan Times, 2008; Singh, 2009). Such an event could have much more devastating impacts than the 1994 GLOF – more than three times the volume of water could gush downhill, an estimated 53 million cubic metres together with rocks and debris (Kuensel, 2010; Singh, 2009). Since the 1994 GLOF, the downstream valleys have become more populated. Many houses and other facilities including important hydropower plants have been built, so such a GLOF event from Thorthormi could have much more serious consequences (Kuensel, 2010; Singh, 2009) with potentially ten times more fatalities and damage (Brauner et al., 2003). Such a precarious situation catalysed the initiation of this project by UNDP.

Key features of the project

As stated by UNDP (2011a), the project aims 'to enhance adaptive capacity to prevent climate change-induced GLOF disasters in Bhutan' with an objective 'to reduce climate-induced risks of GLOFs in the Punakha-Wangdue and Chamkhar Valleys'. Thus, capacity building was a key element of the project and it focused on the downstream areas most at risk from GLOFs from Thorthormi Tsho – the main impact of the 1994 GLOF was in the Punakha-Wandue Valley. During fieldwork as narrated by staff members of agencies involved in the project, and also reported in a number of publications (Bellamy and Penjor, 2014; UNDP 2011a; 2011b, 2011c; UNDP, 2012a; 2012b), there were four main expected outcomes of the project, as discussed below, each with a set of outputs (15 outputs in total, see Bellamy and Penjor, 2014):

- **Outcome 1** – *Improved national, regional and local capacities to prevent climate change-induced GLOF disasters in the Punakha-Wangdue and Chamkhar Valleys*: The Department of Disaster Management (DDM) was responsible for meeting this outcome through building capacity and raising awareness of local government authorities and communities in the project locations. This component of the project was expected to complement the other project activities, including the structural mitigation work in Thorthormi Tsho and particularly the early warning systems. As a consequence of this engagement, DDM developed the Disaster Management Bill that was passed by the parliament and enacted as the Disaster Management Act 2013, mentioned above when discussing policy instruments, a key policy document guiding all disaster-related activities in Bhutan. DDM developed a participatory community-based disaster risk management (CBDRM) training-of-trainers (ToT) manual, which was utilised for training of officers at the *Dzongkhag* level and the formation of *Dzongkhag* DM Committees. These officers then trained personnel at lower administrative levels including *Gewogs* and *Chiwogs*, assisting in the development of DM plans at these levels. Follow-up reviews indicated that staff and community members were strongly aware of the GLOF risk, the early warning systems and evacuation procedures developed in the project (UNDP, 2011b, 2012b).
- **Outcome 2** – *Reduced risks of GLOF from Thorthormi Lake through an artificial lake level management system*: The Department of Geology and Mines (DGM) was the lead agency implementing activities to meet this outcome. Following from extensive scientific research and site investigations, an engineering plan was prepared to lower the water level in Thorthormi Tsho by excavation to widen and deepen its spillway channels. Building on the experience of the *Raphstreng Tsho Outburst Flood Mitigation Project* mentioned earlier in the background to the project, a labour-intensive approach involving about 300 workers was undertaken over four years, using manual equipment (axes, shovels, spades, etc.), drilling machines and silent explosives – to avoid perturbing the moraine and its associated parts. Work could only be done during three to four months each year, generally starting after the monsoons and finishing before the winter season (July to October) to avoid risks that could arise, including downstream flooding and even a GLOF, and also to prevent collapse of the moraine slopes. This required trekking over rugged terrain in very cold conditions for more than ten days to reach the site, carrying equipment on horses, mules and yaks. A helicopter was not available, nor was it considered feasible to procure one. Despite the highly adverse conditions and logistical challenges, the target of lowering the water level by 5 metres was eventually accomplished and more than 17,000

cubic metres of water was drained out – almost equivalent to the amount of water that was discharged in the 1994 Luggye Tsho GLOF. Prior to beginning the project, an Environmental Impact Assessment (EIA) was carried out to identify ways of limiting the project's impact on this protected nature conservation area, and some key strategies were suggested for waste management and firewood collection.

- **Outcome 3** – *Reduced human and material losses in vulnerable communities in the Punakha-Wangdue Valley through GLOF early warning systems (EWS)*: Upgrading the previous manual EWS, the Department of Hydro-Met Services (DHMS) played the key role in implementing an automated EWS in this project; this system has a wider coverage than before. The older system utilised wireless satellite phones (Singh, 2009) and now serves as a back-up. The turn-key installation and supply of equipment was undertaken through a joint venture between USD Enterprises (India) and Sutron Corporation (USA) because of the lack of capacity in this field in Bhutan. At the early stage, training on testing and installation of the equipment was provided by the supplier, enabling DHMS to install the remaining elements of the EWS. Seventeen siren stations and six hydro-met monitoring stations were installed in strategic locations across the two valleys (see Figure 3.7); four of the monitoring stations have Automatic Water Level Sensors to check water levels and the other two are Automated Weather Stations to monitor both water levels and meteorological information. Satellite communication is used to ensure reliability across the long distance between upper and lower regions and can be monitored online. Additionally, GLOF 'red zones' – areas near the rivers posing high risk – were demarcated through hazard zonation mapping to restrict and control development there. The key element to complement the EWS was the identification of evacuation areas in at-risk communities, which typically consisted of an open field on high ground. The automated EWS, though the first of its kind in Bhutan, was contextualised to local conditions. A comprehensive EWS manual was provided by the supplier and DHMS developed a set of Standard Operating Procedures, allowing for the maintenance and effective operation of the system over the long term, as well as providing the opportunity for replication in other areas.
- **Outcome 4** – *Enhanced learning, evaluation and adaptive management*: The project has been extensively documented and its lessons disseminated widely through a number of documentary films broadcast on national television in Bhutan and different international forums. For example, the documentary 'Himalayan Meltdown' was broadcast on *Discovery* and won an award at the Worldfest International Film Festival. A conference was organised for sharing the project's lessons and to develop plans for replication in Bhutan and other countries in the Hindu-Kush mountainous region including Nepal and Pakistan (UNDP, 2012a). Several evaluations of the project were undertaken (for example, Bellamy and Penjor, 2014; UNDP, 2012b) and made available in the public domain.

Figure 3.7 An early warning siren post in Wandue

Disaster risk reduction

Disaster risk reduction, thereby building resilience, was the ultimate purpose of this project focusing specifically on GLOFs. As discussed above in the project's features, a combination of structural and non-structural measures had been undertaken with that purpose in mind. A real-life test of these measures was presented in 2015 when a GLOF occurred after Lemthang Tsho, another glacial lake in the Lunana region, breached its banks due to heavy rainfall during the preceding weeks – rainfall continued incessantly for more than two days (Gurung et al., 2017; Orlove, 2015). The rainfall had alerted downstream district authorities, including in Punakha and Wandue, to the possibility of flooding, and after receiving reports of the GLOF from upstream they activated the EWS sirens manually early in the morning before the water level in the Mochu and other rivers reached a critical threshold, which would trigger the automated system. By the Prime Minister's directive, evacuations were conducted along the Mochu River and were effective within an hour, as both authorities and communities were prepared. An engineer in the Wanduephodrang district early warning control room reported that, 'it was a good opportunity to check that the system actually works' (UNDP, 2016, pp. 44–45). As a local resident narrated, 'The new alarm system is much better than it was during the 1994 event. Only because of that we received early warning and were evacuated.' The efficacy of the EWS implemented in the project was clearly demonstrated in this instance, particularly its combination with an evacuation strategy. Additionally, both institutional capacity and community awareness, key elements that the project aimed to achieve, also became evident.

The Lemthang Tsho GLOF, however, was of much less magnitude than the 1994 Luggye Tsho GLOF. Despite the entire lake becoming empty, the discharge was estimated at 0.37 million cubic metres compared to 18 million cubic metres from Luggye Tsho in 1994 (Gurung et al., 2017). Hence the damage was limited to upstream areas about 30 kilometres from the lake and the GLOF became attenuated by the time it reached downstream areas such as Punakha city. Nonetheless, although there were no human fatalities, four timber bridges were destroyed, an acre of agricultural land was damaged and four horses were killed, among other such impacts (Gurung et al., 2017). This points to the fact that GLOFs continue to pose a threat in Bhutan; the country also faces an uncertain situation of risk with climate change-induced melting of glaciers leading to unpredictable future impacts.

The day the GLOF occurred there was a 5.5 magnitude earthquake about 190 kilometres to the south-east of Lemthang Tsho in the Indian state of Assam (Gurung et al., 2017; Orlove, 2015). It could not be ascertained whether the earthquake contributed to the GLOF, but as discussed above in the section on disaster risk profile of Bhutan, in this seismically active region earthquakes comprise a triggering factor for GLOFs. As evident from the above example of the Lemthang Tsho GLOF, rain-induced GLOFs can usually be predicted because of rainfall patterns and seasonality, but this is not the case for earthquakes. Thus, Bhutan is confronted with a multi-hazard scenario that needs to be taken into account in initiatives for disaster resilience.

National contextuality

This project was highly relevant to the institutional context of Bhutan as it was embedded within NAPA I, a national programme that aimed to address key national priorities including adaptation to the risks posed by climate change. This is a key challenge in a context where retreating and melting glaciers have created a precarious situation with implications for national development and human well-being. It thereby catalysed the establishment of the Disaster Management Act 2013, the most significant national policy instrument for disaster resilience, and consequent formation of the National Disaster Management Authority (NDMA) to formulate and oversee all disaster resilience-related activities in Bhutan. Beyond its framework of expected outcomes of a specific set of outputs, the project charted the course of future policy and action for disaster resilience and had a long-term impact on the governance and institutional landscape of Bhutan. In that sense, the project led to the institutional mainstreaming of disaster risk reduction through a resilience-orientated approach, reflected in the series of subsequent NAPAs.

Although there was important international engagement and financial support – and UNDP was a key partner – the project was entirely led and implemented by national agencies: three government departments (DDM, DGM, DHMS) from two line-ministries (MoEA, MoHCA). This was a valuable opportunity for capacity building within the national governance structure and it also presented the possibility for national ownership and leadership for future disaster resilience endeavours.

Unlike other countries in the region where non-governmental organisations (NGOs) typically play a role in such projects (see for example Ahmed, 2017), here there were no NGOs involved. However, government officials during fieldwork in Nepal acknowledged the potentially valuable role of NGOs and expressed an interest in their future engagement. Thus, this project also allowed for identifying such a potential route for collaboration and partnership to enhance the governmental efforts. This is in line with the 'adaptive management' approach employed in the project, highlighted in a terminal evaluation by Bellamy and Penjor (2014, p. 22). Because of the harsh working conditions in the rugged project site at more than 4,500 metres' elevation, the project team undertook a number of adaptive measures to complete the project within its timeline and budget. Some of the key measures included limiting the structural mitigation work to the spillway channel instead of the entire lake as originally envisioned; provision of additional medical and safety procedures after three workers died from altitude sickness (Business Bhutan, 2010); and involving the Royal Bhutan Army to make up for the deficit of available workers towards the end of the project. Even financial management was found to be adaptive, by 'stretching' every dollar to the extent possible (Bellamy and Penjor, 2014, p. 23). Such an adaptive and flexible approach of governmental agencies augurs well for future undertakings for disaster resilience in Bhutan.

Key challenges

This case study project represents a landmark initiative for disaster resilience. However, significant challenges were experienced during its implementation and also brought into focus broader challenges, as discussed below.

Transport

As mentioned when discussing Outcome 2 above, the project site was in a remote location at a high altitude of more than 4,500 metres without any transport links. Without the availability of a helicopter, the only option was to trek for ten days or so over rugged terrain in harsh and very cold conditions from the nearest head camp, transporting equipment and other project-related materials, as well as supplies of food, water, etc. on animals such as mules, horses and yaks (Nayar, 2009; UNDP 2011a; 2011b; UNDP, 2012b). It was estimated that 1,500 horse (or equivalent) loads per year were required to transport 70–80 tonnes of materials over the 5-year project duration (UNDP, 2012b). This mode necessitated the use of light equipment, overruling heavy machinery or more sophisticated, high-tech equipment as recommended for future projects in several studies (UNDP, 2012a, 2012b).

Similar transport challenges were experienced during the installation of the EWS facilities (Outcome 3) (UNDP, 2011c) rendered by the rugged, hilly topography. However, the distances were not as great as in the case of Thorthormi Tsho and usually reachable in a day or so. The problem of transport, although posing a critical challenge particularly during the structural mitigation work on Thorthormi Tsho, pervades across this mountainous country as discussed above in the section on a country overview of key issues, and constitutes a wider developmental challenge.

A positive spin-off of the animal transport mode, mentioned by several sources (UNDP 2011b; 2011c; UNDP, 2012b), was the local economic benefits to the providers of the animal transport. Additionally, local businesses were able to generate income by selling a range of everyday products and food items – small shops had been built in different areas near to the project site, head camp and key nodes on the transport route, thereby bringing economic benefits to the local communities.

Working conditions

Because of the use of light equipment necessitated by the mode of transport by animals over an extended duration, removing and breaking large boulders to widen the spillway channel presented a significant challenge, making the progress slow. It was not possible to create dry working conditions with the minimal equipment (sandbags, tarpaulin sheets, etc.) and workers had to work in water up to the hip in some places in very cold conditions (UNDP, 2012b) (see Figure 3.8). It was reported that workers fell ill frequently working in such extreme cold conditions, which was exacerbated by poor living conditions such as tarpaulin sheets for tents, thin blankets, monotonous food lacking sufficient nutrition and inadequate working outfits and gear; in contrast, the government officers on site were provided proper tents and sleeping bags (Business Bhutan, 2010; Singh, 2009). Such a situation caused delays and also diminished the enthusiasm of the workforce. Towards the end of the project, despite the payment being higher than such work elsewhere, it became difficult to recruit workers and eventually the Royal Bhutan Army had to be brought in to complete the work (Bellamy and Penjor, 2014; UNDP, 2012b).

Additionally, adverse weather impacted the progress of the work: namely, the heavy rain and flooding brought by Cyclone Aila in 2009, snowfall late in the season in 2011 and a flash flood in 2012 which disrupted access to the site (UNDP, 2012b). In 2010, three workers died from altitude sickness (Business Bhutan, 2010; UNDP, 2012b), further disrupting the morale of the workers. Nonetheless, after these deaths, the medical and safety measures were ramped up: transit medical camps were set up, medical screening of workers and other team members was undertaken, among other such measures, and a medical team was even sent to Switzerland to receive training on altitude sickness (Business Bhutan, 2011; UNDP, 2012b). In the face of the immense difficulties, the project team should be given credit for this adaptive approach to improve the working conditions.

Figure 3.8 Workers wading in icy cold water to carry out manual work in Thorthormi Tsho (courtesy of UNDP-Bhutan)

Residual risk

This project prioritised action on the most threatening glacial lake, Thorthormi Tsho, but the lake is still included in the list of 22 lakes in Bhutan that are considered critical in terms of disaster risk (Gurung et al., 2017). A level of safety has been accomplished by the project, but with ongoing climate change the risk scenario could change. This residual risk was highlighted by the team leader of the project: 'it does not mean there will be no GLOF after the project, even if the water level was reduced by five metres' (Kuensel, 2011). He further observed that an earthquake or large blocks of ice falling from the melting mother glacier might trigger a GLOF (Nayar, 2009). This was encapsulated in a comment by a scientist at the National Centre for Hydrology and Meteorology during fieldwork in Bhutan: 'This is a risk reduction project, not risk elimination.'

Broader issues

The experience of this project allowed Bhutan to gain lessons that brought into focus a range of broader issues relating to disaster resilience, particularly in relation to GLOFs. The remote locations of glacial lakes and barriers to accessibility make it difficult to gather and update data and to monitor the lakes. A participant at the conference on GLOFs supported under Outcome 4 underlined the scantiness of publications in this field (UNDP, 2012a). This is compounded by limited technical capacity, equipment and financial resources. A field technician in Wandue during fieldwork in Bhutan mentioned the need for upgrading the EWS equipment, but there did not seem to be any available funds for that after the project concluded. Nonetheless, beyond this project, there were other initiatives supported by the Japanese government that aimed to address these challenges such as the project 'Study on Glacial Lake Outburst Floods in Bhutan Himalayas' during 2009 to 2012 (Bellamy and Penjor, 2014; Japan International Cooperation Agency, 2013). There is a definite need for such studies over the long term to maintain vigilance on the threat of GLOFs and to monitor and update the existing inventory on glacial lakes.

Human factors also present challenges – for example, when the sirens were activated during the 2015 Lemthang Tsho GLOF (see section above on Disaster Risk Reduction), panic arose in the local communities and they had to be switched off after an hour (Orlove, 2015), yet the CBDRM training developed under Outcome 1 of the project included recommendations for regular mock drills (UNDP, 2012a). Such drills can result in people not taking the sirens seriously over the long term, especially if they are also activated by mistake or due to technical glitches (UNDP, 2012a). Implementing the high-risk red zone demarcation on the ground has also faced hurdles due to human factors (Barrett and Lhamtshok, 2016; UNDP, 2012a). The memory of the 1994 GLOF has faded over time and new residents are not fully aware of the devastation that occurred and therefore do not perceive the risks of living in the red zone.

At the institutional level, particularly with regards to Outcome 1, the sustainability of the capacity building initiatives of DDM might be undermined over the long term due to staff turnover. A lack of a 'permanently designated person' (UNDP, 2012b, p. 47) at DDM and staff at the local authority levels who are responsible for disaster resilience activities over the long term are some of the issues that also affect the sustainability of the project's inputs. Although the project rightly identified capacity building as a key area of focus, the lack of institutional capacity for disaster resilience is a wider challenge in Bhutan, as underscored above when discussing the policy instruments and challenges.

A critical issue pointed out by a participant at the GLOF conference organised under Outcome 4 is that the glaciers and glacial lakes are the main source of water in Bhutan and over the long run if water is continuously drained from them to reduce the risk of GLOFs, it might result in water scarcity (UNDP, 2012a). This could affect a wide range of important sectors including agriculture and hydroelectricity production. An estimated 750 million people in South Asia and China rely on the Himalayan glaciers as their primary source of water (Nayar, 2009); thus, what happens to the glacial lakes in Bhutan have implications for the wider region.

Conclusion

The mountainous country of Bhutan, fringed at its top by one of the greatest mountain ranges in the world – the Himalayas – and its climate change-driven melting glaciers have created conditions for the unique and deadly GLOF hazard. As yet, GLOFs do not occur very frequently, but when they do, as the 1994 Luggye Tsho event demonstrated, they unleash devastating forces with extensive and long-term consequences. Thus, the Bhutan government and its partners rightly prioritised action to reduce the GLOF risk as part of its initial strategies to adapt to climate change.

The hilly and rugged terrain of Bhutan has led to a number of developmental challenges stemming from rapid and unplanned urbanisation, as well as transport difficulties, which have implications for achieving disaster resilience. Despite being a champion in environmental conservation with almost 80 per cent of the country under forest cover, this success is gradually being whittled away by the pressures of urbanisation and population growth; in a way, it is a small-scale microcosm of the transformations evident in other countries of the region.

In the last decade or so, a number of initiatives have been undertaken at the national institutional level to step up policy and practice for achieving disaster resilience, at the roots of which lies the massive 1994 GLOF disaster and the angst it created in its aftermath. Recognition of the link between climate change and such disasters was instrumental in the formulation of a series of NAPAs, the first of which addressed GLOFs. NAPA I was a stepping stone to the long-term commitment of the Bhutanese government and the international development community to undertake integrated climate change adaptation and disaster resilience measures through a combination of structural and non-structural interventions.

The case study project under NAPA I discussed above is the first one of its kind in Bhutan, and one of the most significant initiatives to build resilience against GLOFs. The project took a multi-pronged approach by carrying out structural work to lower the water level in the threatening glacial lake, building capacity for disaster risk reduction at the institutional and community level and installing an automated EWS in the valleys at risk to GLOFs. Notwithstanding the formidable challenges encountered during its implementation, the project largely met its key anticipated outcomes. Because this was such a landmark project, it had relevance beyond Bhutan for the entire Himalayan-Hindu-Kush mountainous region and its replication was initiated in other countries such as Nepal and Pakistan.

Despite the achievements of the project, it managed to reduce, but not eliminate, the risk posed by only one of the 22 dangerous glacial lakes in Bhutan, providing safety to less than 10 per cent of the population. With advancing climate change, whether the project's efforts to prevent icy floods would sustain over the long run – and indeed, with the ongoing melting of glaciers in the region – the future prospects for disaster resilience for the whole nation remain uncertain. Events in Bhutan have transboundary implications for the vast populations in nearby countries relying on the glaciers for their water. But the project relays a simple message: such a mammoth problem cannot be addressed in its entirety, but a concerted effort to do so, such as this project, is a hopeful sign.

References

Ageta, Y. and Iwata, S. (1999) *The Assessment of Glacier Lake Outburst Flood (GLOF) in Bhutan*. Nagoya, Tokyo and Thimphu: Institute of Hydrospheric-Atmospheric Sciences of Nagoya University, Department of Geography of Tokyo Metropolitan University and Geological Survey of Bhutan.

Ahmed, I. (2017) 'A Partnership-Based Community Engagement Approach to Recovery of Flood-Affected Communities in Bangladesh' in Marsh, G., Ahmed, I., Donovan, J. and Barton, S. (eds.) *Community Engagement in Post-Disaster Recovery*. Abingdon (UK): Routledge, pp. 22–36.

Awale, S. (2019) *Weather Warning for Climate Meeting*. Nepalitimes.com. Available at: www.nepalitimes.com/banner/weather-warning-for-climate-meeting/?fbclid=lwAR07z4WGX3_yUa98ef0CoKcilzuABf5fYwnRVH_MH8y-0XT (accessed 9 August 2019).

Barrett, A. and Lhamtshok, T. (2016) *Bhutan: Disaster Risk Management Status Review*. Thimphu: Department of Disaster Management (DDM).

Bellamy, J.J. and Penjor, Y. (2014) *Terminal Evaluation: Reducing Climate Change-Induced Risks and Vulnerabilities from Glacial Lake Outburst Floods in the punakha, Wangdue and Chamkhar Valleys (GLOF)*. Thimphu: United Nations Development Programme (UNDP).

Bhutan Times. (2008) 'Taming Thorthormi'. *Bhutan Times*, 19 November 2008.

Bouissou, J. (2011) *As Glaciers Melt, Bhutan Faces Increased Risk of 'Mountain Tsunamis'*. Le Monde/Worldcrunch. Available at: http://content.time.com/time/world/article/0,8599,2098960,00.html (accessed 25 September 2019).

Brauner, M., Leber, D., Häusler, H., Payer, T. and Agner, P. (2003) *Glacier Lake Outburst Flood (GLOF) Mitigation Project, Lunana, Bhutan: Technical Mitigation Measures, Thorthormi Outlet*. Vienna: University of Vienna.

Build Change. (2016) *Bhutan*. Available at: www.buildchange.org/locations/bhutan/ (accessed 8 July 2019).

Business Bhutan. (2009) 'Water Level Lowered by 86 Cm'. *Business Bhutan*, 28 October 2009.

Business Bhutan. (2010) 'Thorthormi's Sacrificial Lambs'. *Business Bhutan*, 11 September 2010.

Business Bhutan. (2011) 'Almost a Quarter of Bhutanese Glaciers Have Shrunk in Last Three Decades'. *Business Bhutan*, 13 December 2011.

ChhimidD Consulting. (2018) *An Assessment of Post-Project Scenario in Terms of national, Local and Community Capacity to Prepare for and Respond to Climate-Induced Multi-Hazards*. Thimphu: National Environment Commission Secretariat (NECS), Royal Government of Bhutan (RGoB) and United Nations Development Programme (UNDP).

Civil Intelligence Agency (CIA). (2018) *The World Factbook: Bhutan*. Available at: https://www.cia.gov/library/publications/the-world-factbook/geos/bt.html (accessed 6 February 2020).

Climate Council. (2017) *Bhutan Is the World's Only Carbon Negative Country, so How Did They Do It?* Available at: www.climatecouncil.org.au/bhutan-is-the-world-s-only-carbon-negative-country-so-how-did-they-do-it/ (accessed 26 September 2019).

Dema, K. (2016) *International Tourists Bemoan Bad Road Conditions in Bhutan*. Asia News Network. Available at: web.archive.org/web/20160708150611/http:/www.asianews.network/content/international-tourists-bemoan-bad-road-conditions-bhutan-18438 (accessed 11 July 2019).

Dema, T. (2018) *Landslides at Ossey Close Traffic for 18 days*. KuenselOnline. Available at: www.kuenselonline.com/landslides-at-ossey-close-traffic-for-18-days/ (accessed 11 July 2019).

Department of Engineering Services. (2017) *Windstorm Resilient Roofing System* (Guideline). Thimphu: Ministry of Works & Human Settlements.

Department of Human Settlement. (2017) *Building Regulations & Land Pooling Regulations*. Thimphu: Ministry of Works & Human Settlements.

Dorji, U. (2017) 'Enhancing Preparedness for Glacial Lake Outburst Floods' in Wymann von Dach, S., Bachmann, F., Alcántara-Ayala, I., Fuchs, S., Keiler, M., Mishra, A. and Sötz, E. (eds.) *Safer Lives and Livelihoods in Mountains: Making the Sendai Framework for Disaster Risk Reduction Work for Sustainable Mountain Development*. Bern (Switzerland): University of Bern and Bern Open Publishing, pp. 58–59.

Druk Green Power Corporation. (2019) *Basochhu Hydropower Plant*. DrukGreen. Available at: www.drukgreen.bt/basochhu-hydropower-plant/ (accessed 24 September 2019).

Ebrahim, Z.T. (2019) *As Glacial Lake Bursts, Preparation Saves Lives in Pakistan*. PreventionWeb. Available at: www.preventionweb.net/news/view/66587?&a=email&utm_source=pw_email (accessed 9 August 2019).

Fujita, K., Nishimura, K., Komori, J., Iwata, S., Ukita, J., Tadano, T. and Koike, T. (2012) 'Outline of Research Project on Glacial Lake Outburst Floods in the Bhutan Himalayas'. *Global Environmental Research*, 16, pp. 2–12.

Geological Survey of Bhutan. (1994) *Preliminary Report on the Investigation of Glacial Lakes at Pho Chhu Source and the Assessment of Flood Affected Areas in Lunana*. Thimphu: Geological Survey of Bhutan.

Gerretsen, I. (2019) *Himalayan Glaciers are Melting Twice as Fast as Last Century*. CNN. Available at: https://edition.cnn.com/2019/06/19/world/himalayan-glaciers-melting-climate-change-scn-intl/index.html?no-st=1561076118 (accessed 24 September 2019).

Goldenberg, S. (2011) *Glacier Lakes: Growing Danger Zones in the Himalayas*. The Guardian. Available at: www.theguardian.com/environment/2011/oct/10/glacier-lakes-melt-himalayas (accessed 11 July 2019).

Gurung, D.R., Khanal, N.R., Bajracharya, S.R., Tsering, K., Joshi, S., Tshering, P., Chhetri, L.K., Lotay, Y. and Penjor, T. (2017) *Lemthang Tsho Glacial Lake Outburst Flood (GLOF) in Bhutan: Cause and Impact*. Lalitpur (Nepal): International Centre for Integrated Mountain Development (ICIMOD).

Gyelmo, D. (2016) *Forest Fires Burning up Bhutan's Wilderness*. The Third Pole. Available at: www.thethirdpole.net/en/2016/03/21/forest-fires-burning-up-bhutans-wilderness/ (accessed 11 July 2019).

Gyelmo, D. (2018) *As Bhutan's Economy Grows, so Does Its Waste Problem*. The Third Pole. Available at: www.thethirdpole.net/en/2018/04/06/as-bhutans-economy-grows-so-does-its-waste-problem/ (accessed 11 July 2019).

Häusler, H., Leber, D., Schreilechner, M., Morawetz, R., Lentz, H., Skuk, S., Meyer, M., Janda, C. and Burgschwaiger, E. (2000) *Final Report of Raphstreng Tsho Outburst Flood Mitigatory Project (Lunana; Northwestern Bhutan): Phase II*. Vienna: University of Vienna.

Himalayas. (n.d.) *Local Transportation in Bhutan*. Explore Himalayas. Available at: www.himalaya2000.com/bhutan/travel-tips/local-transportation.html (accessed 26 September 2019).

International Life Saving Federation. (2009) *Cyclone Aila Relief and Rescue*. Available at: www.ilsf.org/2009/06/24/cyclone-aila-relief-and-rescue/ (accessed 11 July 2019).

Ives, J.D., Shrestha, R.B. and Mool, P.K. (2010) *Formation of Glacial Lakes in the Hindu Kush-Himalayas and GLOF Risk Assessment*. Kathmandu, Geneva and Washington, DC: International Centre for Integrated Mountain Development (ICIMOD), United Nations Office for Disaster Risk Reduction (UNISDR) and Global Facility for Disaster Risk Reduction (GFDRR).

Japan International Cooperation Agency (JICA). (2013) *With JICA's Help, Bhutan Establishing a System for Early Warning on Floods Caused by Global Warming*. JICA. Available at: www.jica.go.jp/english/news/field/2013/131113_02.html (accessed 26 September 2019).

Kattelmann, R. (2003) 'Glacial Lake Outburst Floods in the Nepal Himalaya: A Manageable Hazard?'. *Natural Hazards*, 28, pp. 145–154.

Komori, J., Koike, T., Yamanokuchi, T. and Tshering, P. (2012) 'Glacial Lake Outburst Events in the Bhutan Himalayas'. *Global Environmental Research*, 16, pp. 59–70.

Kuensel. (2010) '1.3 M down after 3-Month Stint'. *Kuensel*, 7 October 2010.

Kuensel. (2011) 'Battle Done, War Not Over'. *Kuensel*, 12 November 2011.

Kuensel. (2012a) 'Glacial Districts in Bhutan'. *Kuensel*, 1 January 2012.

Kuensel. (2012b) 'Last Nu 33M for Last Lap'. *Kuensel*, 29 February 2012.

Kuensel. (2016a) *North-East-West Highway Widening Progressing Under Pre-Financing*. KuenselOnline. Available at: www.kuenselonline.com/north-east-highway-widening-progressing-under-pre-financing/ (accessed 11 July 2019).

Kuensel. (2016b) *East-West Highway Widening on Track*. KuenselOnline. Available at: www.kuenselonline.com/east-west-highway-widening-on-track/ (accessed 11 July. 2019).

Lamsang, T. (2009) *Flood Kills 4 (Bhutan)*. ICIMOD. Available at: www.icimod.org/?q=5197 (accessed 11 July 2019).

Lhamo, P. (2018) *NAPA II Project Identifies Four Critical Landslides and Flood Prone Areas*. KuenselOnline. Available at: www.kuenselonline.com/napa-ii-project-identifies-four-critical-landslides-and-flood-prone-areas/ (accessed 11 July 2019).

Lotay, Y. (2015) *Country Report: Bhutan Disaster Management*. Kobe (Japan): Asian Disaster Reduction Center (ADRC).

Marti, R.Z. (2005) *The 2004 Hurricanes in the Caribbean and the Tsunami in the Indian Ocean*. New York: United Nations.

Meenawat, H. and Sovacool, B.K. (2011) *Improving Adaptive Capacity and Resilience in Bhutan*. Singapore: Springer Science + Business Media B.V.

Mellino, C. (2016) *This Country Isn't Just Carbon Neutral ... It's Carbon Negative*. EcoWatch. Available at: www.ecowatch.com/this-country-isnt-just-carbon-neutral-its-carbon-negative-1882195367.html (accessed 11 July 2019).

Mool, P.K., Wangda, D., Bajracharya, S.R., Kunzang, K., Gurung, D.R. and Joshi, S.P. (2001) *Inventory of Glaciers, Glacial Lakes and Glacial Lake Outburst Floods Monitoring and Early Warning Systems in the Hindu Kush-Himalayan Region*. Kathmandu: International Centre for Integrated Mountain Development (ICIMOD).

Morelle, R. (2019) *Spy Satellites Reveal Extent of Himalayan Glacier Loss*. BBC News. Available at: www.bbc.com/news/science-environment-48696023 (accessed 24 September 2019).

National Assembly of Bhutan. (2013) *Disaster Management Act of Bhutan 2013*. Thimphu: National Assembly of Bhutan.

National Environment Commission (NEC). (2016) *Bhutan State of the Environment Report 2016*. NEC.

National Statistics Bureau of Bhutan. (2006) *2005 Population & Housing Census of Bhutan*. Thimphu: National Statistics Bureau of Bhutan.

National Statistics Bureau of Bhutan. (2018) *2017 Population & Housing Census of Bhutan*. Thimphu: National Statistics Bureau of Bhutan.

Nayar, A. (2009) *When the Ice Melts*. Nature.com. Available at: www.nature.com/news/2009/091021/full/4611042a.html (accessed 20 August 2019).

Norbu, T. (2012) *Disaster Management System in Bhutan* (PowerPoint presentation). Thimphu: Department of Disaster Management (DDM).

Orlove, B. (2015) *Glacier Lake Bursts in Bhutan*. GlacierHub. Available at: https://glacierhub.org/2015/07/01/glacier-lake-bursts-in-bhutan/ (accessed 11 July 2019).

Oxford Poverty and Human Development Initiative. (n.d.) *Bhutan's Gross National Happiness Index*. Available at: https://ophi.org.uk/policy/national-policy/gross-national-happiness-index/ (accessed 24 September 2019).

Parliament of Bhutan. (2009) *The Local Government Act of Bhutan 2009*. Thimphu: Parliament of Bhutan.

Pashley, A. (2015) *Bhutan: Climate Lessons from a Himalayan Kingdom*. Climate Home News. Available at: www.climatechangenews.com/2015/08/05/bhutan-climate-lessons-from-a-himalayan-kingdom/ (accessed 11 July 2019).

Penjore, T. and Sato, G. (2012) *Landslides in Bhutan. Kuensel: Environment *17*. Available at: www.jica.go.jp/project/bhutan/002/kuensel/pdf/kuesnsel_20120331.pdf (accessed 26 September 2019).

Phuntsho, S. (2018) *Road Traffic Accidents Down by 40 Per Cent*. BBS. Available at: www.bbs.bt/news/?p=88610 (accessed 11 July 2019).

The Poultry Site. (2016) *Bird Flu Hits Backyard Poultry in Bhutan*. Available at: https://thepoultrysite.com/news/2016/10/bird-flu-hits-backyard-poultry-in-bhutan (accessed 11 July 2019).

Pradhan, B. (2014) *Disaster Risk Management: Bhutanese Scenario* (PowerPoint presentation). Thimphu: National Land Commission Bhutan.

Preventionweb. (2014). *Bhutan – Disaster & Risk Profile*. Available at: www.preventionweb.net/countries/btn/data (accessed 9 August 2019).

Rai, R. (2018) *Landslide Closes Manitar-Lhamoizingkha Road*. KuenselOnline. Available at: www.kuenselonline.com/landslide-closes-manitar-lhamoizingkha-road/ (accessed 11 July 2019).

Reuters. (2018) *Bhutan Reports Outbreak of Severe H5N1 Bird Flu: OIE*. Available at: www.reuters.com/article/us-health-birdflu-bhutan/bhutan-reports-outbreak-of-severe-h5n1-bird-flu-oie-idUSKCN1GP237 (accessed 11 July 2019).

Royal Government of Bhutan (RGoB) and United Nations Development Programme (UNDP). (2014) *Addressing the Risks of Climate-Induced Disasters through Enhanced National and Local Capacity for Effective Actions* (NAPA II project document). Thimphu: RGoB and UNDP.

Royal Government of Bhutan (RGoB), World Bank (WB) and the United Nations (UN). (2009) *Joint Rapid Assessment for recovery, Reconstruction and Risk Reduction*. Thimphu: RGoB, WB and UN.

Royal Government of Bhutan (RGoB), World Bank (WB) and the United Nations (UN). (2011) *Joint Rapid Assessment for recovery, Reconstruction and Risk Reduction*. Thimphu: RGoB, WB and UN.

Singh, S.M. (2009) *The Cost of Climate Change the Story of Thorthormi Glacial Lake in Bhutan*. Thimphu: World Wide Fund for Nature (WWF).

Trigona, C. (2018) *Bhutan Road Network*. Dlca.logcluster.org. Available at: https://dlca.logcluster.org/display/public/DLCA/2.3+Bhutan+Road+Network (accessed 26 September 2019).

Tshering, T. (2017) *State of Bhutan's Hydropower Projects*. Yesheydorji.blogspot.com. Available at: http://yesheydorji.blogspot.com/2017/06/state-of-bhutans-hydropower-projects.html (accessed 11 July 2019).

Tutton, M. and Scott, K. (2018) *What Tiny Bhutan Can Teach the World About Being Carbon Negative*. CNN. Available at: https://edition.cnn.com/2018/10/11/asia/bhutan-carbon-negative/index.html (accessed 11 July 2019).

United Nations Development Programme (UNDP). (2007) *Reducing Climate Change-Induced Risks and Vulnerabilities from Glacial Lake Outburst Floods in the Punakha-Wangdi and Chamkhar Valleys*. Thimphu: UNDP.

United Nations Development Programme (UNDP). (2011a) *Reducing Climate Change-induced Risks and Vulnerabilities from Glacial Lake Outburst Floods in the Punakha-Wangdue and Chamkar Valleys* (project factsheet). Thimphu: UNDP.

United Nations Development Programme (UNDP). (2011b) *Reduced Risks of GLOF from Thorthormi Lake through an Artificial Lake Level Management System* (outcome 2 factsheet). Thimphu: UNDP.

United Nations Development Programme (UNDP). (2011c) *Reduced Human and Material Losses in Vulnerable Communities in the Punakha-Wangdue Valley through GLOF Early Warnings* (outcome 3 factsheet). Thimphu: UNDP.

United Nations Development Programme (UNDP). (2012a) *Glacial Lake Outburst Flood (GLOF)*: Reducing Risk and Ensuring Preparedness (conference proceedings). Paro (Bhutan): UNDP.

United Nations Development Programme (UNDP). (2012b) *Reducing Climate Change-induced Risks and Vulnerabilities from Glacial Lake Outburst Floods in the Punakha, Wangdue and Chamkar Valleys*. Thimphu: UNDP.

United Nations Development Programme (UNDP). (2016) *Adapting to the Himalayan Meltdown*. Thimphu: UNDP.

United Nations Development Programme (UNDP). (2017) *Addressing the Risks of Climate-induced Disasters through Enhanced National and Local Capacity for Effective Actions*. Thimphu: UNDP.

United Nations Development Programme (UNDP). (n.d.) *UNDP: Bhutan's Partner for Happiness* (leaflet). Thimphu: UNDP.

United Nations Environment Programme (UNEP). (2009) *Recent Trends in Melting glaciers, Tropospheric Temperatures over the Himalayas and Summer Monsoon Rainfall over India*. Nairobi: UNEP.

USD Enterprises, Sutron Corporation. (2010) *Inception Report for Installation of Glacial Lake Outburst Flood (GLOF) Early Warning System in the Punakha-Wangdi Valley*. Thimphu and Sterling (VA, USA): USD Enterprises and Sutron Corporation.

Wangchuk, T. (2012) *Disaster Management in Bhutan*. Thimphu: Department of Disaster Management (DDM).

Wangdi, C., Gurung, M.S., Duba, T., Wilkinson, E., Tun, Z.M. and Tripathy, J.P. (2018) 'Burden, Pattern and Causes of Road Traffic Accidents in Bhutan, 2013–2014: A Police Record Review'. *International Journal of Injury Control and Safety Promotion*, 25(1), pp. 65–69.

Wangdi, N. and Kusters, K. (2012) *The Costs of Adaptation in Punakha, Bhutan: Loss and Damage Associated with Changing Monsoon Patterns*. Bumthang (Bhutan) and Amsterdam: Ugyen Wangchuck Institute for Conservation and Environment and Wereld in Woorden – Global Research and Reporting.

Watanabe, T. and Rothacher, D. (1996) *The 1994 Lugge Tsho Glacial Lake Outburst Flood, Bhutan Himalaya*. JSTOR. Available at: www.jstor.org/stable/3673897?seq=1#page_scan_tab_contents (accessed 20 August 2019).

World Bank. (n.d.) *Regional Program for Hydromet Services and Resilience* (South Asia: program brief). Washington, DC: World Bank.

World Population Review. (2019a) *World Population by Country*. Available at: http://worldpopulationreview.com/ (accessed 26 September 2019).

World Population Review. (2019b). *Bhutan Population 2019*. Available at: http://worldpopulationreview.com/countries/bhutan-population (accessed 8 July 2019).

World Population Review. (2019c) *Kathmandu Population 2019*. Available at: http://worldpopulationreview.com/world-cities/kathmandu-population/ (accessed 11 July 2019).

The 2004 Indian Ocean Tsunami

Disaster recovery in the Maldives

4 Staying afloat

Disaster recovery in the climate change-threatened atolls of the Maldives

Introduction

This chapter explores disaster resilience aspects in the Maldives, drawing from the recovery activities undertaken after the 2004 Indian Ocean Tsunami, the largest disaster to ever affect the country. This island nation was severely impacted by the tsunami. As the Maldives consists of low-lying scattered atolls, it is highly vulnerable to such disasters, and is at extreme risk of climate change impacts such as sea level rise. For the Maldives, staying afloat could become a serious challenge in the future. Unlike other tsunami-affected countries – where recovery projects have been extensively studied – the research on tsunami recovery in the Maldives is limited. For example, a key publication edited by Shaw (2014) on the long-term aspects of recovery from the tsunami includes case studies from a number of tsunami-affected countries, but not the Maldives. This chapter addresses that gap.

Post-disaster recovery is often a protracted process with intrinsic linkages to a multitude of sectors. Achievements and shortcomings become evident after a long period of time – after newly built settlements are inhabited and communities become established. It is therefore relevant to look at recovery in the Maldives at this point in time – more than a decade after the disaster.

There are two main components of this chapter. Firstly, an overview is presented of the context, including: the impacts of the tsunami; relevant critical issues in the Maldives; its disaster risk profile; and policy framework for disaster resilience. Secondly, an analysis of a case study in Dhuvaafaru island – one of the largest recovery projects in the Maldives – is provided. The first component delineates the opportunities and challenges for achieving disaster resilience in the country; the second component, the case study, provides an understanding of how disaster resilience initiatives are applied in practice via a recovery project. This is pertinent in relating to the post-tsunami 'build back better' ethos that has become globally prevalent among the disaster resilience community of practice and research (Fernandez and Ahmed, 2019). Through analysis of each case study in this book, the intention is to understand how actions within a specific case study context may elicit wider resonance.

The research for this chapter involved two key methodological approaches. Firstly, a wide range of relevant literature was reviewed; secondly, fieldwork for data collection was undertaken in the Maldives, including the case study island of Dhuvaafaru. The main data collection methods were stakeholder and key informant interviews, community meetings, transect walks, observation, and photographic and audio-visual recordings. The data from the literature review and fieldwork was then analysed through content analysis, which led to grouping of themes, coding and cross-referencing to interpret and present the findings of the analysis below.

Disaster overview: 2004 Indian Ocean Tsunami

The 2004 Indian Ocean earthquake and tsunami that struck on the morning of Boxing Day, 26 December, was one of the most massive disasters in recent history, with wide-ranging impacts. Waves, in some places more than 30 metres high, devastated coastal communities in 14 countries and killed almost 230,000 people (New World Encyclopedia, 2018). Aftershocks were felt as far away as Alaska (Asian Development Bank, 2005). The impacts of the tsunami were extensive, with maximum damage experienced by the housing sector, as noted in Indonesia – the most severely affected country (Marti, 2005).

The tsunami originated near the coast of Sumatra, Indonesia, and travelled rapidly at 700–800 kilometres per hour. It reached the Maldives around 9:20am, about three and a half hours after the tsunami struck Indonesia (Saleem and Sattar, 2009; World Bank, Asian Development Bank and UN System, 2005). Although pandemonium had already broken out in Indonesia, due to the absence of an early warning system in the Maldives, the population was caught unawares. Also, as being a public holiday, most people in the Maldives had been relaxing that morning and had not tuned in to the media; nobody knew about the chaos that was occurring in the wider region.

As the Maldives consists of dispersed low-lying atolls in the Indian Ocean, the tsunami's 1.2–4.2 metre waves swept across the country; 82 people were killed, more than 1,300 were injured and 26 persons went missing and were feared to be dead. The waves also caused extensive destruction to property and infrastructure. This was a disaster of a national proportion, affecting more than 80 per cent of the 198 islands comprising the country, a third of its nearly 290,000 population at that time, and nearly 10 per cent of the population displaced from their islands or homes. Some displaced people had to be accommodated in temporary camps for more than a decade – until the end of 2015. Fourteen islands out of the 198 inhabited ones were completely devastated; their residents had to be evacuated. Fifty-six islands experienced significant physical damage and 121 were affected to some degree by the inundation. As elsewhere, the tsunami had a strong impact on the housing sector with more than 2,500 houses destroyed, more than 3,500 houses severely damaged and another 1,500 houses requiring repair. Electricity and telecommunications links were widely disrupted for several months, and water supply was interrupted in about 15 per cent of the islands. Jetties and harbours providing connections to the capital Malé and to other islands were affected in about 25 per cent of the islands – a significant impact that resulted in people being marooned without essential services and supplies (tsunami impact sources: Ministry of Planning and National Development, 2005; Saleem and Sattar, 2009; Risk Management Solutions India, 2006; United Nations Development Programme, 2006; United Nations Environment Programme, 2005; WB, ADB and UNS, 2005; interviews of organisational staff in the Maldives).

Compared to other tsunami-affected countries, the death toll in the Maldives was much lower. Where Indonesia suffered around 170,000 fatalities, and Sri Lanka suffered around 36,000 (New World Encyclopedia, 2018), the Maldives was protected because of the unique geophysical feature of scattered islands (Fritz et al., 2006). A relatively lower population and lesser ferocity of the tsunami also resulted in lower fatalities. However, in the Maldives, much more economic damage was experienced than in other tsunami-affected countries: 62 per cent of the national GDP was lost, compared to 2.6 per cent in Indonesia and 4.5 per cent in Sri Lanka (UNDP, 2006). Additionally, the impact of the tsunami in the Maldives was more widespread, with almost all the 199 inhabited islands affected (MoPND, 2005). Two of the mainstays of the national economy – tourism and fisheries – were severely impacted (Saleem and Sattar, 2009; WB, ADB and UNS, 2005). Before the tsunami, tourism contributed to nearly 70 per cent of the Maldives' GDP (Saleem and Sattar, 2009); following the tsunami, drastic reduction in tourist numbers continued for several years, affecting the livelihoods of the large number of people working in this sector. Fisheries providing livelihoods to a significant proportion of the population were also severely affected – not only because of the loss of fishing equipment and infrastructure, but also due to the displacement of many people from their homes. Agriculture in the Maldives was also affected, as inundation by saline water led to the loss of an annual agricultural harvest for almost 15,000 farmers (RMSI, 2006).

Country overview of key issues

The Maldives, officially the Republic of Maldives, consists of 1,190 low-lying coral islands which sit about 1–1.5 metres above sea level. The island clusters, or atolls, which make up this small South Asian country are scattered across 900 kilometres of the Indian Ocean (see Figure 4.1) (Ministry of Planning and National Development, 2005; RMSI, 2006). It is the smallest country in Asia in terms of population (World Population Review, 2019). Key critical issues of national significance relating to disaster resilience are discussed below.

Population density

At the time of the 2004 Indian Ocean Tsunami, the Maldives' population of nearly 290,000 (countryeconomy.com, 2004) was widely disbursed across the islands, with concentrations in urban centres – particularly in the capital island of Malé. According to WB, ADB and UNS (2005) about 198 islands were inhabited and a third of them had a population of less than 500; about 70 per cent of the islands had less than 1,000 people living on them. Only three of the inhabited

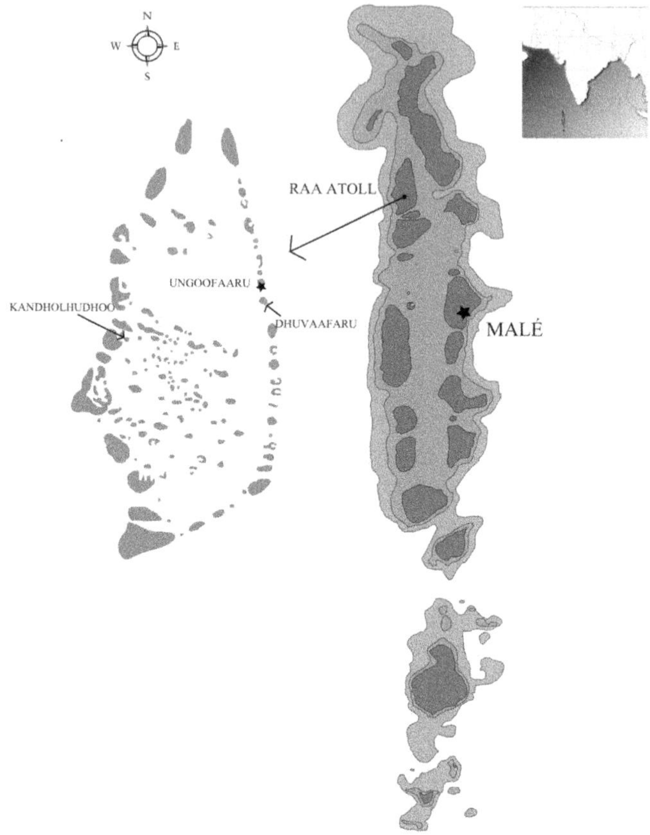

Figure 4.1 Map of the Maldives showing locations relevant to this chapter

islands had an area of more than 4 square kilometres; nine islands had more than 2 square kilometres (Ministry of Fisheries and Agriculture, n.d.) and 28 islands had an area of more than 1 square kilometre (WB, ADB and UNS, 2005). Despite the small sizes of the islands, at the time of the 2004 tsunami the Maldives had a high population density: 963 persons per square kilometre, on average (countryeconomy.com, 2004). By 2018, 14 years after the tsunami, the population had risen to more than 445,000 – nearly a 54 per cent increment. Consequently, the population density has also increased, estimated at 1,700 persons per square kilometre, making the Maldives the eighth most densely populated country in the world (World Population Review, 2019).

Urbanisation

Despite the Maldives' dispersed islands setting, it is a rapidly urbanising country with nearly 45 per cent urban and more than 4.5 per cent annual urban growth

(PreventionWeb, 2014); growth is mostly concentrated in the capital, Malé. A key issue for the Maldives is the primacy of Malé: its population of more than 100,000 is estimated to make up nearly 20 per cent of the national population (National Bureau of Statistics, 2014; World Population Review, 2019). The next largest city, Addu, has less than one-third of Malé's population. Even the top 20 cities of the Maldives have significantly lower populations; some of the smaller cities have populations of only 500–1,000 inhabitants (World Population Review, 2019). Malé is among the world's most densely populated cities (World Population Review, 2019). With so many people concentrated in its 5.8 square kilometres (Zahir, 2011a), there is very little scope for expansion, other than land reclamation and the construction of high-rise buildings – both of these practices have long-term negative environmental consequences.

The Malé urban jurisdiction is comprised of six divisions, two of which are separate islands. The main island of Malé is only 1.77 square kilometres (Hello Publishing, 2018) and has more than 90,000 residents, with a density of more than 47,000 persons per square kilometre (National Bureau of Statistics, 2006). The island is very densely built up, with hardly any green areas; it has many buildings over ten stories high, with more developments in planning or under construction at present (see Figure 4.2). The tallest building in construction at the time of writing is 25 stories (vNews, 2016). It is not clear to what extent

Figure 4.2 Malé island is a densely built-up city

such a small coral island can continue to sustain such heavy and dense construction. Accommodating the growing population – including extensive migration from other Maldives islands – will be a key issue for Malé in future.

Even in the other larger islands, there are limited livelihood opportunities outside of the tourism industry and fisheries. This is a key factor in propelling migration to Malé for work in other service sectors, along with the need for better options for education, health care and other facilities. Indeed, as Faisal (2008) observed, this phenomenon has resulted in depopulation of the lowly populated smaller islands and excessive urbanisation of Malé. Thus, while the communities in the smaller islands are fragmenting, and local capacity and human resources are being drained, heavy pressure is being placed on services, infrastructure and employment opportunities in Malé (WB, ADB and UN, 2005). Basic services such as water supply, sanitation and waste management present critical challenges; the amount of solid waste littering the streets and disposed of indiscriminately into the surrounding ocean poses a serious environmental concern in Malé.

Transport

The primacy of Malé poses significant problems for maritime transport of people, goods and services to the dispersed islands (RMSI, 2006). There is hardly any industrial development in the Maldives and almost all commodities are imported (National Disaster Management Centre, 2010). Commodities pass through Malé and are then transported to the other islands. This usually entails long distances by boat, involving significant time and money, as well as embodied fossil fuel energy. Thus, basic commodities tend to be expensive in the Maldives compared to other South Asian countries. On the other hand, the country has sustained a high economic growth rate of about 7 per cent in recent years, with a per capita GDP of more than US$10,000 (World Bank, 2018) – much higher than the other South Asian countries. This contributes towards a purchasing power to match the high commodity costs. However, it is uncertain to what extent an economy based mainly on tourism and fisheries can be sustained in the future – especially in the face of climate change, coral bleaching, declining fish stocks and overfishing (Afzal et al., 2016; Cheung et al., 2013). Tourism is likely to be affected by future climate change impacts and disasters, as seen after the 2004 tsunami (RMSI, 2006). These developments could potentially impact the livelihoods of a large segment of the population, undermining their resilience through significant shocks and stresses. Despite the strong national economy, economic vulnerability prevails widely – particularly in Malé, with its high costs of housing, utilities and food (Faisal, 2008).

The geography of the Maldives' scattered islands necessitates widespread passenger movement, as well as the transport of commodities. This leads to extensive fuel consumption – mainly diesel – and again, expenditure of time and money. Some of the smaller islands do not have the necessary facilities for health care and education, posing a significant risk during medical emergencies

(RMSI, 2006), and restricting the availability of basic health care services and inoculation. Residents of the smaller islands must travel to the larger islands to access education – particularly higher education – and to attend work at the resort islands. This regular movement of people requires extensive transport. Ferry services are limited in the outer islands and often do not function very well, and private boats are expensive (RMSI, 2006). For longer distances, it is necessary to fly – again, flights are expensive and generally unaffordable for the wider population. Many of the outer islands face serious socio-economic developmental challenges because of this limited accessibility (NDMC, 2010).

Workforce

Another key issue in the Maldives was the lack of sufficient local workforce. More than 15 per cent of Maldivian residents are foreigners (NBS, 2014) – a significant proportion of the population. In many cases, Maldivians are reluctant to work in manual labour-based jobs (Lawther, 2010), and there is a steady flow of foreign labour, largely from other Asian countries – particularly poorer ones such as Bangladesh, India, Nepal and Sri Lanka; this workforce consists of both documented and undocumented workers (Burke and Rasheed, 2015; Majeed, 2018). Tensions between locals and foreigners are common, adding to other common problems such as very low wages and pressure on undocumented workers from local authorities (Burke and Rasheed, 2015; Mohammed, 2015). Thus, questions arise about whether this reliance on foreign workers can be sustained over the long term, and whether local occupational capacity building might offer a better alternative.

Disaster risk profile

The Maldives does not have a history of recurrent and high intensity natural hazards and is considered as having moderate disaster risk (NDMC, 2010; RMSI, 2006); the 2004 Indian Ocean Tsunami was a sudden and isolated disaster originating from a faraway source. However, the massive impact of this event brought national awareness, leading to the establishment, under a presidential decree, of the National Disaster Management Centre (NDMC) (MoPND, 2005; Moosa, 2014; NDMC, 2010). Many of the disaster risk assessment studies in the Maldives were undertaken subsequently, allowing a more comprehensive understanding of the country's disaster risk profile (see for example RMSI, 2006).

As evidenced by the 2004 tsunami, the Maldives' national economy is at risk from disaster events and climate stresses, and the tourism- and fisheries-dependent economy is highly sensitive in this regard. Among all the islands, Malé has been assessed as having the highest disaster risk. The fast pace of 4.5 per cent urban growth (PreventionWeb, 2014) compounds an already complex disaster risk scenario, and requires a different set of resilience-building approaches for the increasingly depopulated outer islands and densifying urban centres. The Maldives will be required to plan and deliver services that both improve the quality of life for residents on the different islands and make the growing cities resilient to natural hazards in future.

Since the 2004 tsunami, a range of hazards have been identified in the Maldives. The following main hazards – posing disaster risk and diminishing potential for resilience – have been garnered from key sources including Hay (2006); Moosa (2014), NDMC (2010), RMSI (2006) and United Nations Framework Convention on Climate Change (2001); other sources with more specific information have been cited where relevant.

Tsunamis and earthquakes

Despite the fact that the Maldives is an archipelago with extensive coastal exposure, the 2004 Indian Ocean Tsunami was the first significant tsunami to hit the islands; this had devastating consequences, as discussed above. As characterised by RMSI (2006, p. 16), 'low probability and high consequential tsunami hazard in the near future' poses a risk. It was estimated that the 2004 tsunami has a return period of 219 years, which means that another tsunami equal to or of more magnitude could occur at least once within a 219-year period; in each year, there is a 1 per cent probability of a tsunami of such magnitude to occur (for a simplified explanation of a return period see Hay, 2006). Due to the location of the Sumatra Subduction Zone to the east – the origin of the 2004 tsunami – the eastern Maldivian islands and the eastern coasts are more at risk of tsunamis. Other seismic faults in the region to the west in the Indian Ocean and Arabian Sea also pose a risk, though to a lesser extent. The southern islands have been assessed to be at risk of strong earthquakes, with potential for light to moderate damage, but this has not yet occurred. Nonetheless, mild tremors continue to be experienced – particularly from the east, originating in the Sumatra fault (Maldives Independent, 2016; TripAdvisor, 2012). Hence, despite the lack of historical earthquake impact, there is a risk. A powerful earthquake in densely built-up Malé could be highly destructive, especially considering that the national building code has very little provision for seismic resilience (see Ministry of Construction and Public Infrastructure, 2008).

Windstorms

Windstorms in the Maldives include both localised thunderstorms and cyclones which form over the ocean and impact over a wide area. They include strong wind, heavy rainfall, lightning, and sea surges – all hazards in themselves. Recent cyclones in the Indian Ocean and the cyclone-prone Bay of Bengal – such as Cyclone Nilam in 2012 (World Health Organization, 2012) and Cyclone Ockhi in 2017 (Maldives Independent, 2017) – caused extensive impacts on many of the northern islands, with inundation by surges and damage to properties. However, compared to other countries in the region affected by these cyclones, such as India and Sri Lanka, the impacts on the Maldives are less severe. The Maldives is not considered a cyclone-prone country, and it is mainly the northern atolls that are at risk. Nonetheless, the vulnerability of these atolls is compounded by their remoteness and difficulty of access for providing emergency services. With climate change, windstorms are expected to increase in intensity; stronger wind gusts are expected, and thus the existing moderate risk could magnify. Seasonal localised storms and windstorms with heavy rainfall and lightning are more widespread in the Maldives now, and often cause damage to property and infrastructure as well as disruption to work and activities.

Climate change impacts

The low-lying atolls of the Maldives are exposed to anthropogenic global warming-induced climate change impacts; it is one of the most threatened countries in the world. With an average land height of 1–1.5 metres above sea level, or even less in some islands, an estimated 49–95 centimetre rise in sea level by the end of this century is expected to inundate vast parts of the archipelago. Heavy inundation due to sea level rise is anticipated to occur at least annually by 2050. It has been estimated that 15 per cent of Malé will be inundated by 2025, and 31 per cent by 2050, resulting in total inundation by the end of this century (RMSI, 2006; UNFCCC, 2001). These are significant projections, and if even a proportion of them eventuates, it would spell trouble for the densely populated and built-up city; as Malé is the national hub, this would have repercussions for the entire country. At present, Malé island is surrounded by a seawall – this provided a level of protection during the 2004 tsunami (Zahir, 2011a), and may ward off the rising seawaters in the short term. However, during fieldwork it was observed that it does not seem to be well maintained, with cracks and peeling plaster evident in many places. Also, whether its height and strength are sufficient to cope with future sea level rise thresholds is uncertain.

Therefore, institutional initiatives seek to promote both disaster risk reduction and climate change adaptation (NDMC, 2010), and to understand the linkages between disaster and climate change risks. Increases of intensity and incidence of extreme weather events is a key link between climate change and disasters (Intergovernmental Panel on Climate Change, 2012), but other negative impacts such as heatwaves, salinity intrusion, coastal erosion and rising sea levels can reduce resilience and add to the vulnerability of the Maldivian island communities. All climate change impacts are expected to amplify over time due to global warming. Yet, when discussing this with island communities, many people remarked that they have no idea about how climate change might impact them, and have not experienced any impacts so far. Clearly, in an area that is so seriously vulnerable to the effects of climate change, there is a need for raising awareness about climate change and its effects.

Drought and water shortage

Another serious problem in the Maldives is seasonal drought, resulting in a shortage of water. This occurs from November to April, particularly in the northern atolls. A key source of water – subsoil freshwater lenses, or aquifers at shallow depths – has already been over extracted and contaminated by both human waste and salinity. Before the 2004 tsunami, only a limited number of 39 inhabited islands had drinkable groundwater (UNEP, 2005); the tsunami further aggravated this problem by the intrusion of saltwater. Future climate change is expected to result in 'overwash' (Storlazzi et al., 2018, p. 1) – that is, overtopping of banks and ingress of waves inland due to a higher sea level, leading to salinity seepage into groundwater. Rainwater tanks are widely used in the Maldives, but they were often not adequate to tide people through the dry season. Another approach to the issue of drought has been the construction of desalinisation plants. Through the NDMC and the Ministry of Environment and Energy (MoEE), the government provides emergency water supplies to almost 50 drought-affected islands. However, this comes at a high cost – in some recent years more than US$500,000 had been spent annually on supplying water (Moosa, 2014). This is compounded by the cost of embodied energy, time and resources expended in transporting the water. Desalinisation plants, particularly in outer island locations, are costly investments and have high operation and maintenance costs. Some Island Councils do not have a budget for this and face severe water scarcity in the dry season (Hamid, 2015). The United Nations Development Programme (UNDP) is working with the MoEE to seek options to address climate change-induced water shortages in the future.

Coastal erosion

Perennial coastal or beach erosion is widespread in the Maldives, with an estimated 97 per cent of the islands being affected (Mangroves for the Future, n.d.). While the island setting provides a natural exposure to sea waves, human activities such as building harbours and jetties, dredging and land reclamation, construction of coastal settlements, and clearing coastal vegetation such as mangroves have exacerbated the problem. Extraction of coral from the reefs around the islands has reduced the protective capacity of the reefs, allowing stronger waves to whittle away at the beaches (Zahir, 2011b). After the 2004 tsunami, clean-up operations involved cutting down coastal mangroves, which further exposed the islands to erosion (UNEP, 2005). With a rising sea level due to climate change, the erosion hazard is expected to become accelerated, accompanied by salinity intrusion both inland and into groundwater aquifers. Loss of land and eroded coastal settlements comprise a significant hazard, affecting the key sectors of tourism and fisheries. The seriousness of the hazard has stimulated governmental and non-governmental action to undertake construction of protective seawalls, plantation of coastal vegetation and restriction of coral mining; the efficacy of these measures will only be understood over time, and it is unclear whether the damage already done will be reversible.

Floods

Large sea swells are a threat in this country of low-lying atolls and have resulted in coastal flooding on several occasions, such as the heavy floods of 1987 and

2017; these events are linked to climate change-induced sea level rise and other factors such as tidal conditions, and are anticipated to increase in the future with climate change (Wadey et al., 2017). Flooding due to heavy rainfall is a localised hazard, and will be discussed further in the Dhuvaafaru project case study. The extensively built-up urban areas and lack of sufficient green areas with permeable surfaces in cities such as Malé have affected the natural hydrology and water run-off, leaving them vulnerable to flooding and waterlogging.

Human-induced hazards

The geographical setting of dispersed islands in the Maldives entails a large volume of maritime traffic; consequently water accidents of different types – from capsizes and collisions, to missing boats and drowning – are all too common (see for example the inventory of accidents of the Maldives Times, 2019). The level of safety protocols is not clear, and it does seem that there is scope for risk reduction. Other key human-induced hazards include accidents on construction sites, explosions and fires. The NDMC listed seven fire incidents in 2014 (NDMC, 2014) causing damage to property and infrastructure. The shortage of water in the dry season, as mentioned above, presents a constraint in fire-fighting, and because of the scattered nature of the locations of the islands, it is difficult to provide timely and adequate emergency services. Another key human-induced hazard in the Maldives is the deficiency of solid waste management (Saleem and Sattar, 2009). It is common to see significant volumes of litter constantly washing up on the beaches of the different islands. In urban centres, such as in Malé, solid waste littering is widespread on streets and in surrounding water, especially near harbours. This environmental hazard is out of control, and there seems to be very little public awareness or provision for solid waste management.

Policy aspects of disaster resilience

As pointed out by a staff member of UNDP, 'The tsunami was a wake-up call leading to establishing NDMC and the evolution of the DRR [disaster risk reduction] framework in the country.' The policy and institutional landscape in the Maldives relating to disaster resilience centres on disaster risk reduction and climate change adaptation; this is because of the cognisance of disaster risk generated by the 2004 tsunami, and also the high threat posed by climate change, as discussed below.

Administrative structure

The government administrative structure of the Maldives operates at two main levels: national and local. All national government agencies are based in Malé. The oversight of local government agencies is provided by the Local Government Authority (LGA) under the Ministry of Home Affairs. The local government operates at three levels: City Councils for islands with more than 25,000 people, of which there are two – Malé and Addu; Atoll Councils in 20 administrative atolls; and Island Councils on all the inhabited islands – except those with City Councils – with elected members. These Island Councils are administered and monitored by their respective Atoll Councils (Moosa, 2014).

Institutional arrangement for disaster resilience

At the national level, the NDMC, under the Ministry of Defence and National Security, is the prime body responsible for all activities relating to disasters, including emergency management, recovery and disaster resilience initiatives, and coordination during disasters. Together with the Maldives National Defence Force (MNDF) and the Maldives Police Force (MPF), NDMC leads emergency response operations. Prior to the 2004 Indian Ocean Tsunami, the Maldives was considered to have low disaster risk, and hence there was no institutional structure for disaster risk management and reduction. In emergencies, the MNDF undertook response operations, which is perhaps why NDMC was created under the aforementioned ministry after the tsunami disaster. Reportedly, in response to recurring sea swells and flooding, there was discussion at the government level more than a decade before the tsunami to establish a disaster management centre and committee, but the decision to actually form such a body was catalysed by the devastation wreaked by the tsunami. It is unfortunate that often it takes a disaster to initiate resilience initiatives – this is common throughout the world.

NDMC now works in coordination with the LGA to support Disaster Management Units at City, Atoll and Island Councils. These units are provided oversight by Disaster Management Standing Committees at respective levels. Some of the main government bodies that NDMC collaborates with include the Department of Meteorology – for early warnings – and the Environment Protection Agency, to address environmental hazards. NDMC works closely with relevant non-governmental organisations such as the CARE Society and the Maldives Youth Climate Network, United Nations agencies including UNDP and UNICEF, and the Maldivian Red Crescent Society. During the 2004 tsunami recovery, NDMC worked in coordination with the International Federation of Red Cross and Red Crescent Societies (IFRC) and different national Red Cross/Crescent societies, as will be discussed below in the Dhuvaafaru project case study.

Policy instruments and challenges

Some of the key policy instruments relating to disaster resilience include:

- *National Emergency Operation Plan* (NEOP), which was approved in 2018. NEOP focuses largely on emergency response, and related roles and coordination mechanisms.
- *National Disaster Management Plan*, which is being developed. It draws from a provisional draft of the *Strategic National Action Plan for Disaster*

Risk Reduction and Climate Change Adaptation 2010–2020, where the necessity of synergy between actions to address climate change and disasters is emphasised (NDMC, 2010).

- *Disaster Management Act*, which was ratified in 2015. This act articulates the responsibility of the government to take a lead role in activities relating to disasters, in line with the Constitution of the Maldives.
- *National Adaptation Programme of Action* (NAPA) to address climate change was initiated before the 2004 tsunami and subsequently updated. Eventually the NAPA transformed into the *Strategic National Action Plan on Climate Change Adaptation and Disaster Risk Reduction for 2010–2020*, which clearly articulates the intrinsic link between climate change and disasters in the Maldives.

The above initiatives indicate a strong national policy interest in developing disaster resilience, catalysed no doubt by the sudden and unprecedented massive tsunami disaster. However, it is not clear to what extent this interest is being translated into tangible actions to enable resilience building. Despite all these policy efforts, as pointed out by Moosa (2014), NDMC has very few major projects and only a very limited budget that covers operational and staff costs; funds for emergency response are available when required, not for undertaking long-term disaster resilience programs. The Maldives is still perceived as a low-risk country, and as mentioned by an NDMC official, 'The priority given to disaster management at the policy level is a bit low.'

Most of the disaster resilience projects are funded by external agencies, and they tend to be modest in scope. For example, a key project supported by UNDP on *Strengthening School Preparedness for Tsunamis in Asia and the Pacific* in 18 countries – where the Maldives is one of the five priority countries – mainly focused on safety drills, first aid training and awareness of fire and electrical hazards (Oaku, 2018). While the initiative was praiseworthy, given that there are 198 scattered inhabited islands, it was only possible to run a single drill or training event on a limited number of islands. This does not allow for the kind of wide-ranging and repeated coverage required to reinforce learning, nor does it allow for any follow-up work to monitor the impact and continuity of new safety practices.

Most initiatives like this tend to focus on training undertakings, and other than in Malé and some isolated examples, there are hardly any that support physical resilience and a wider gamut of activities required to address the different hazards and critical problems affecting the Maldives. As in the above example of the UNDP project, a project on *Community Based Disaster Risk Management* mentioned as 'successful' by an NDMC official focused largely on training activities, as did most of its projects (NDMC, 2016). NDMC has attempted to implement actions to the extent possible, given its limited budget; however, there is still scope for assessing how a wider range of initiatives could be translated on the ground in this highly threatened island country. Some work in this regard has been initiated recently – again with the support of UNDP –

that went beyond training and undertook innovative work, such as using drones for risk mapping of the different islands, and methodology for assessing damage and loss in disasters. At this point, however, the future trajectory of disaster resilience in the Maldives does not seem clearly defined.

Safe Islands Programme

A key initiative relevant to this chapter is the *Safe Islands Programme* (SIP) of the government (Suppasri et al., 2015). Prior to the 2004 tsunami, under the *Sixth National Development Plan 2001–2005*, a plan was formulated through the *National Population Consolidation Strategy and Programme* to relocate communities to demarcated 'focus islands' (WB, ADB and UNS, 2005, p. 8). At that time, the main idea was to consolidate population clusters to achieve economic efficiency in transport and delivery of services such as health care and education. After the tsunami, with the new impetus for disaster resilience and also climate change adaptation, these islands were planned to be established as resilient settlements under the SIP – focused around locations that were relatively less exposed to maritime hazards, and incorporating resilience features such as seawalls, elevated land and evacuation areas (MoPND, 2005). This plan was also a land use management strategy (Riyaz and Park, 2010). The Dhuvaafaru case study project discussed below is an outcome of the SIP.

The geophysical setting of the low-lying and small, scattered islands of the Maldives presents significant challenges in achieving a high level of disaster resilience, and in that sense the SIP is limited in scope to low-magnitude hazards. As observed in an assessment by UNDP (2008, p. 21), 'none of the islands in the Maldives is safe from natural hazards.' The low elevation of the islands and lack of high ground for evacuation makes it very difficult, if not impossible, to provide safety from high-magnitude flooding events, due to sea swells and surges; with a rising sea level, this might become even more difficult.

An important critique of the SIP is that consolidating a greater percentage of the population into small islands exposes more people to hazards – localised disasters could have more severe impacts (NDMC, 2010; UNDP, 2008). Additionally, a higher density settlement places more pressure on natural resources and infrastructural development; this could increase disaster risk and lead to a wide range of hazards such as flooding, waterlogging and land erosion, as discussed above in the disaster profile of the Maldives.

Case study: Raa Dhuvaafaru post-tsunami recovery project

Raa Dhuvaafaru, hereinafter referred to as Dhuvaafaru, was previously an uninhabited island in the northern Raa Atoll of the Maldives. It had an area of about 0.48 square kilometres (48 hectares), elongated in the north–south direction, with its length about three times the width (NBS, 2018) (see Figure 4.3). After the community of the nearby island of Kandholhudhoo was devastated in the 2004 Indian Ocean Tsunami, they were settled on Dhuvaafaru island as part of the disaster

Figure 4.3 An aerial view of Dhuvaafaru during construction of the project (courtesy of NDMC)

recovery. Dhuvaafaru is about five times the size of Kanholhudhoo. The Dhuvaafaru 'safe island' project was a completely new settlement – designed, planned and built after the tsunami under the government's *Safe Island Programme* (SIP) (see previous section on policy arrangements for disaster resilience).

The project was implemented by the International Federation of Red Cross and Red Crescent Societies (IFRC) in partnership with NDMC, and some other Red Cross/Crescent partners. Within NDMC, a temporary Housing and Infrastructure Reconstruction Unit was set up for recovery operations across the different islands, which undertook the work for the project including planning and designing the project, particularly the housing and infrastructure. The bulk of the construction was undertaken as a joint venture between the Singapore-based Lian Beng Group and the Maldivian Amin Construction; Lian Beng played a more significant role, indicating the limited national capacity for such projects. The discussion below has been garnered from interviews with associated agency staff and others living in Dhuvaafaru, supplemented by relevant literature.

Pre-tsunami conditions in Kandholhudhoo

Before the tsunami, Kandholhudhoo was a densely populated island with more than 3,600 people living on around 7 hectares of land, with an average density of more than 9,000 persons per square kilometre; this was extended by reclamation to around 11 hectares (Naylor, 2015). The reclamation had negative consequences – by filling in the water between the original island and the reef around it, the island's natural protection was reduced, leading to greater coastal exposure to sea swells and flooding. The central part of the island was at a lower height, and thus experienced waterlogging after heavy rainfall. To avoid the congestion on the island, many houses were built on the reclaimed land, only a few metres from the sea, and were thus highly exposed to coastal inundation (Bluepeace, 2009). The reclamation rendered the island's surface impermeable, reducing the replenishment of its freshwater lens; as such, the local water wells – the main source of drinking water – became contaminated (Naylor, 2015). With no sewerage system on the island, sewage was discharged directly into the surrounding sea, causing local pollution.

The densely congested island had small houses clustered closely together, with very narrow walkways between them that were prone to becoming waterlogged. In such an environment, clashes between neighbours, gang fights, domestic violence, child abuse and other such anti-social behaviour was rampant and highly visible. It was reported that in such small houses, there was sometimes no privacy from children for sexual activity. There was no scope for expansion; at the same time, the island's population was growing.

The government proposed moving people to the uninhabited island of Dhuvaafaru on the same Raa Atoll under its *National Population Consolidation Strategy and Programme* (see section above on policy arrangements for disaster resilience), but this was initially rejected. People had strong ties to the island because of their primary livelihood in fisheries; the open sea on the island's west allowed ample fishing opportunities. In fact, despite the congested living conditions in Kandholhudhoo, many households there were wealthy from the income made from fishing and fish or seafood processing. However, for the growing population of young people, conditions on the island were unbearable, with no sports facilities or playgrounds and limited educational, recreational and employment opportunities. In 2003, youth representatives approached the government and initiated a survey to assess willingness for relocation; they were able to eventually convince more than 90 per cent of the people on the island to agree to relocate. Before they could relocate, however, the tsunami hit, devastating the island; after this, the relocation initiative became the only option for this community. This was not necessarily viewed negatively by community members. As commented by a social worker from Kandholhudhoo, 'The tsunami was a positive thing that happened for the community, it wasn't really a disaster.'

Impacts and aftermath of the 2004 Indian Ocean Tsunami

Kandholhudhoo was one of the most severely affected islands in Raa Atoll and, indeed, in the Maldives (International Federation of Red Cross and Red Crescent Societies, n.d). Most of the houses there were destroyed, and the entire island was rendered uninhabitable: sea water contaminated its freshwater lens and wells, annihilated the vegetation, destroyed the fishing and fish processing equipment, and consequently disrupted the main livelihood of people there.

The tsunami struck Kandholhudhoo around 9:30am on a clear and sunny Sunday morning on 26 December 2004. It was a public holiday and people were relaxing when they were suddenly met with rapidly rising water, which quickly engulfed the whole island. Unlike other affected countries such as Indonesia, Sri Lanka and Thailand, where massive waves battered the coast, Kandholhudhoo experienced a sudden onrush of water which rose to around two to three metres, flooding the houses and roads. Some people reported that they unexpectedly found themselves swimming in deep water. Many of the island's residents thought it was a tidal swell, which is common; no one had ever heard of a tsunami. Many people were injured, and three people died – two women and one man, two of whom were elderly. The elderly man had refused to escape to safety, reportedly giving up to fatalism, saying, 'The end of the world is coming, no need to move.' His attitude is indicative of the unprecedented and abrupt nature of the event and the fear that it can bring to a wholly unprepared community.

During the crisis, youth volunteers helped people move to their roofs or the local multi-storey school and mosque, where they took refuge for four to five hours in the ensuing chaos. The Atoll Council from the atoll capital, Ungoofaaru island, then sent boats to rescue the people in Kandholhudhoo; the whole island was then evacuated and abandoned. More than 3,000 displaced people were initially accommodated in public buildings and by host families in Ungoofaaru and nearby islands. After this, with the support of IFRC the government built transitional shelters in five islands: Ungoofaaru (more than 60 per cent stayed here); Hulhudhuffaaru; Alifushi; Meedhoo; and Maduvvaree (IFRC, 2010). The population of Kandholhudhoo stayed on these islands for more than four years while the new settlement in Dhuvaafaru was constructed.

It was reported by IFRC (n.d.) that the transitional shelters were built even better than the internationally recognised Sphere minimum standards for disaster response (Sphere Project, 2004). An NDMC official mentioned that there were very few complaints regarding the transitional accommodation, and sometimes the shelters were better than the housing of the host communities on the islands. All of the residents' day-to-day needs, such as electricity and water, were taken care of with generous donations from many international organisations; even furniture and cooking implements were provided, which many of the disaster victims took with them later to the new housing in Dhuvaafaru. People began reviving their fishing livelihoods and even started small shops on the host islands; some of the residents were even able to begin saving money, since most necessities were provided. This was confirmed in discussions with Dhuvaafaru residents, who were generally satisfied with the transitional provisions; despite the modest size of the shelters, they accepted it with the understanding that it was a temporary arrangement. While it took more than four years before relocating to Dhuvaafaru – and almost ten years before all the basic facilities there were completed, due to various factors causing delay, as discussed below – the displaced people were content. In fact, after moving to Dhuvaafaru, when residents were informed that from now they would have to pay for services such as electricity, water and waste disposal, many were shocked and unhappy, as they had become used to everything being provided for them free of charge. For that reason, each household was given a start-up grant of approximately US $4,000 to make the transition to self-reliance and permanent housing easier. Even costs of transport and removal of belongings were covered by the government and IFRC.

Key features of Dhuvaafaru 'Safe Island Project'

The Dhuvaaafaru recovery project was the largest such project ever done before by the IFRC (IFRC, 2008); 600 new houses, out of which 562 were contributed by IFRC and 38 by the government, were completed on Dhuvaafaru towards the end of 2008 when people started moving there. The settlement was officially inaugurated by the Maldivian President at that time in March 2009. While people began living and establishing themselves in the new settlement, many other facilities, some of them essential, were built and continued to be added to for many years after. Even during the field visit for this chapter in 2018, footpaths and drains (see Figure 4.5), and a new fish market were being built, and roads were planned to be paved. Another 76 houses near the harbour and jetty were built by the government from 2010 to 2011. The final cost of facilities funded by IFRC and its partners was approximately US$43 million (IFRC, 2010). Figure 4.4 shows the central part of Dhuvaafaru on a scale model with some of the key facilities including the Island Council building. Some of the main facilities on this newly planned and built settlement include:

- 676 houses (562 funded by IFRC, 45 by the government). All houses had a 2500-litre rainwater tank; additional larger tanks were provided in the public buildings. The 86 square metre three-bedroom houses were single-storey and built on 185.4 square metre plots.
- An Island Council building, with the Island Court, Island Disaster Management Unit, an ATM and additional government offices.
- Three schools (preschool, primary and secondary).
- Two mosques: one Friday mosque, built by the government, and a community mosque built with funding from Saudi Arabia.
- A health centre, provided by the German Red Cross.
- A community hall, with a multipurpose auditorium, seminar rooms and a library.
- Parks and recreational areas, including a FIFA-sized football ground, recreational park area and neighbourhood parks. The site plan for the island included green buffer areas to separate different zones. Additionally, 500 trees were planted as part of Environment Day in 2008.
- A jetty and harbour.
- A fish market, which was under construction as of 2018.
- Private commercial developments, including shops, retail outlets, and cafés or restaurants.
- A light industrial zone, with fish processing facilities, water bottling and ice production plants.

Figure 4.4 The central part of Dhuvaafaru on a scale model

- A power station with electricity distribution network, including photovoltaic solar power panels. This was integrated within a utility compound under the Malé Water and Sewerage Company (MWSC).
- A desalination plant built later in 2013 and maintained by MWSC.
- A solid waste management centre with incinerator. Waste is periodically taken by boat to a large waste management centre in Tilafushi island. The residents paid MWSC for this service.
- Sewerage, including septic tanks in each house. It has been reported that effluent is discharged into the sea with maximum dilution and minimal

environmental impact (IFRC, 2010). MWSC ran the sewerage system under its corporate social responsibility programme.

- A road network and street lighting, with footpaths and drains under construction in 2018; (see Figure 4.5).
- A telecommunications tower, built later in 2013 by private service providers, which provides good internet and mobile phone service on the island.

It is clear from the above that a wide range of facilities were provided to make Dhuvaafaru a habitable settlement. It had the essential services and community infrastructure to support the housing, unlike many recovery projects in other tsunami-affected countries where only houses with very basic infrastructure were built (Ahmed and McEvoy, 2014). It was a definitive improvement from the previous living conditions of the pre-tsunami community in Kandholhudhoo, described above.

Disaster risk reduction

IFRC conceptualised this project as a 'build back better' initiative (IFRC, 2008, 2010, p. 4), where not only resilient houses were built, but an attempt was made

Figure 4.5 Island Council building in Dhuvaafaru serves as a safe refuge. Note on the bottom left the start of footpath construction

to build all the community infrastructure and services to a better standard than pre-tsunami conditions. After the tsunami, the community expressed their desire to remain in Raa Atoll, where their livelihoods were based, and also to stay together because of the strong kinship ties within the community. The decision for relocation to Dhuvaafaru was based on earlier plans before the 2004 tsunami, under the government's consolidation strategy and the ongoing social and infrastructure problems that the community in Kandholhudhoo faced, as previously described. Other factors in the decision to relocate included the ready availability of an uninhabited island of a sufficient size and elevation on the atoll. That the reef around Dhuvaafaru also provides a degree of protection from tidal swells and surges was also a decisive factor to relocate in order to reduce the disaster risk to the community (IFRC, 2010).

Other disaster resilience aspects of the Dhuvaafaru project include the Island Council building, which is centrally located on the island and elevated on pillars to serve as a safe refuge during possible coastal hazard impacts (see Figure 4.5). The council level Disaster Management Unit was located in this building, providing disaster risk reduction and emergency management support to the community. The local drills and training provided through this unit also raised community awareness. As narrated by a resident of Dhuvaafaru, 'Before we knew very little about disaster preparedness, but now we are much more aware.' The tsunami has had a long-term psychological impact, embedding fear in the community; every time an alert was issued, or even if there was an unrelated loud noise, anxiety arose within the community. Such post-tsunami anxiety was also reported elsewhere in other rehabilitated islands in the Maldives (Suppasri et al., 2015). During the drills facilitated by UNDP, the fearful memories of the tsunami surfaced, and many participants felt hesitant to go through the drill (Oaku, 2018). Nonetheless, in the new settlement there was a greater sense of security – residents took comfort in the proximity of large buildings such as the Island Council, schools, mosques and Heath Centre, the spacious layout and drainage which allows water to flow away from the island, and the placement of houses, which are set back further from the coast. These new structures and infrastructure improvements all contribute to disaster resilience.

One of the main problems in Dhuvaafaru is perennial coastal erosion. The western side – which faces the lagoon, harbour and jetty – is protected by a seawall, but the eastern and southern coastlines are largely exposed; the light industry zone is particularly at risk. Parts of the eastern side still retain a portion of the original mangroves, which provide some protection; in other places, the Maldives National Defence Force has helped to build some concrete barriers. However, erosion is a long-term process and is expected to be exacerbated in future due to sea level rise. Despite the efforts to establish the settlement in a relatively elevated island, Dhuvaafaru is only about a metre above sea level. It was reported that the original plan was to raise the level of the land via landfill, but this part of the plan was not carried out, in the end, because of the extra cost. In fact, during the settlement construction process, most of the vegetation on the island was removed and the land levelled by removing topsoil, resulting in the island losing some of its original elevation and protection (Bluepeace, 2009; Saleem and Sattar, 2009; Talbot, 2012). Some residents were unhappy that they were promised a 'safe island', when actions such as this were contrary to that promise. In many flood-prone contexts elsewhere, houses are built on a high plinth or raised on stilts (Ahmed, 2017), but in the Maldives it was reported that this was not a common practice, and introducing such measures might be unacceptable because of lack of precedent.

Several sources highlight the risk faced by Dhuvaafaru and note that, at present, it only has resilience to low-magnitude coastal hazards (Bluepeace, 2009; Naylor, 2015). As with most islands in the Maldives, Dhuvaafaru is exposed to coastal hazards and long-term impacts of climate change; it is exceedingly difficult to reduce this risk beyond a certain degree. However, it was reported by a Maldivian Red Crescent agency staff member that in some other islands, raised land has been created to provide better protection – this is something that the Maldives will need to consider over the long term.

One of the main issues facing the Dhuvaafaru community is their limited awareness of climate change. One resident mentioned, 'I don't feel any climate change. But this island is very open and hotter, previously [in Kandholhudhoo] there was more shade in the narrow lanes.' It seems that the experience of the tsunami has eclipsed the concern about other hazards. While at the institutional level there was significant preoccupation with the challenges of climate change, this has not percolated to the community level, even though such communities face a significant risk.

Nonetheless, if viewed from a relative standpoint, the community in Dhuvaafaru is in a much better situation in terms of resilience compared to when it was in Kandholhudhoo. Dhuvaafaru does not experience the kind of flooding and waterlogging problems that were common in Kandholhudhoo; additionally, Dhuvaafaru has a drainage system, a protective reef and safe refuges with institutional support for disaster risk reduction, together with a range of other provisions. The houses are much sturdier, with earthquake-resistant construction (Oughton, 2009) and the community infrastructure has been built to a reasonable quality. As captured succinctly in a comment by a Dhuvaafaru resident, 'This island is not safe, but safer than before.'

Key challenges

Notwithstanding the significant achievements of the Dhuvaafaru project, there were a number of challenges during its implementation and after it was completed, as discussed below.

Community engagement

A key issue was the limited community consultation and participation in the project (Saleem and Sattar, 2009), which has led to a number of problems. Even a staff member of IFRC commented that, 'One area where we could have improved upon was around consultation with the communities and involving them to a greater degree in the project while the island was under construction' (Talbot, 2012). There were hardly any staff members in IFRC with the right skills to facilitate community participation (Saleem and Sattar, 2009), which was also a problem in other projects implemented by its partner societies in the Maldives (Lawther, 2010). During the implementation of the project in Dhuvafaaru, most of the community resided in transitional camps elsewhere. There was, however, some effort to involve the community: initial consultation meetings confirmed the desire of the community to relocate to Dhuvaafaru based on the recommendation of NDMC (Arup International Development, 2012; IFRC, 2010). Additionally, a small team provided updates to the community and sometimes arranged site visits; this team provided subsequent training to community members to maintain the facilities on the island after construction was completed (IFRC, n.d.; Lawther, 2010).

On the other hand, it was conveyed by IFRC that, 'Working with the community of R. Kandholhudoo was reported to be a difficult task' (Saleem and Sattar, 2009, p. 32). The community was characterised as 'demanding and proud', and expected payment for any involvement, signifying a level of dependency. Perhaps because everything had been paid for after the tsunami – as mentioned above when discussing the tsunami's impact and aftermath – this was a natural outcome of the dependent position that the community had been placed. A strategy for community engagement and ownership of the project's planning and process right from the outset might have obviated or at least reduced such dependency. Nonetheless, it should be acknowledged that in the aftermath of such a massive disaster, and with a project of this proportion, an agency with limited experience of working in such a situation would face considerable barriers to implementing community engagement. The fact that the original community had been scattered on five different islands during the project construction was also not conducive to implementing a concerted strategy for community engagement.

Beneficiary allocation

Another issue stemming from the limited community engagement was the extensive disputes and disagreements that arose in the allocation of housing. Even though this was undertaken through a fair and transparent lottery process, some people felt they had been disadvantaged. Being accustomed to living in extended families, large families wanted houses close together; unfortunately, the lottery process did not always allow for this, which led to disagreements.

A key issue in the allocation of beneficiaries was determining who was a resident of Kandholhudoo. At the time of the tsunami, there were some people who belonged to Kandholhudoo but did not reside there; hence, they were not allocated houses in Dhuvaafaru. Some belonged to large extended families who lived together in Kandholhudoo, but as the houses on Dhuvaafaru were smaller than the community's original dwellings – built to accommodate no more than eight persons – some of these non-resident family members were left out, leading to grievances. To address these complaints, the government built a further 76 houses (see section above on the project features). However, the disagreements persisted, resulting in subsequent denial of registration by the government; this is further discussed below in the sub-sections on long-term issues and transformations.

Project implementation

The project experienced significant delays, for a number of reasons. Negotiations at all stages between the government and IFRC were lengthy, worsened by a change in government in 2008; at one point the project funding was planned to be diverted by the government to other activities, requiring extensive dialogues (Saleem and Sattar, 2009). There were also complications in the tendering process – while IFRC preferred a single contractor, it eventually became a joint venture between Lian Beng Group, Singapore and Amin Construction, Maldives; this was because most Maldivian construction firms did not have the capacity for a project of this size. Furthermore, the bidding and contracting process had many complications, including allegations of corruption. Costs also increased in the interim, causing revision of budgets and contract documents, adding to the delay. By the time the project commenced, it was mid-2006, a year and a half after the 2004 tsunami (IFRC, 2010).

As mentioned in the above discussion of low levels of industrial production and key transportation issues in the Maldives, almost all the building materials for this project had to be imported, and the geography necessitated transportation of these goods by boat over long maritime distances (IFRC, 2010; Saleem and Sattar, 2009). Even local materials such as sand and aggregates proved difficult to secure locally. These problems led to several extensions to the contract, significantly affecting the project's implementation.

Insufficient and inadequate human resources at different levels also affected the project (O'Donnell and Kakande, 2007). Local skilled management staff were scarce, particularly because of a high staff turnover in agencies. Although it was recommended to use local construction workers, it proved more cost-effective to employ migrant workers from other countries in the region – even from as far as the Philippines. The large bulk of the workers were from Bangladesh, a country which generally supplies a significant proportion of the low-paid work force in the Maldives (for example see Majeed, 2018); many Bangladeshi workers settled afterwards in Dhuvaafaru, running restaurants and cafés, working in the fish processing industry, and staying in separate housing built for them by the fish processing firms.

A positive factor reported by Saleem and Sattar (2009) was the role played by women in the project, through their participation in meetings and focus group discussions. However, as the Maldives is a predominantly male-dominated Muslim country, this meant that the extent of such participation was limited, and final decisions were usually made by the men.

Construction drawbacks

Although there was an effort by IFRC to build the houses to a high standard, in most of the houses many defects became evident after construction. Cracks appeared in different parts of the house, door locks did not work, sanitary fixtures were defective, water penetrated through vents under the roof, connections between different parts of the house were inadequate and incorrect roofing screws and low-quality roofing sheets had been used. There were defects in some of the public facilities as well; for example, problems with the sewers and mismatching tiles in the school buildings. Nonetheless, there was a Defects Liability Period under which most of these problems were rectified, except for the most difficult ones. For example, it was not considered feasible to replace the roofing sheets of all the 600 houses (IFRC, 2010). Thus, while beneficiaries were generally satisfied after the repairs, they were compelled to live with the unresolved defects.

Some of the construction problems were of a contextual nature. For example, vents under the roof of the houses and schools were designed with the good intention of providing ventilation for thermal relief in the warm climate, but it was not understood that in the exposed landscape of the island, monsoonal rain could be driven horizontally by strong wind, allowing water to penetrate to the interior of the buildings (IFRC, 2010). In this coastal environment, the vents also allowed sand to be blown into the houses. Over time, people took their own measures to close these vents. Adequate community consultation on the design of houses and public facilities could have prevented such a waste of resources of both the implementing agencies and beneficiaries.

Unfortunately, the additional 76 houses built by the government (see section above on the project's features) were reportedly of lower quality than the IFRC houses. Their foundations were weaker and did not allow for construction of an extra storey; cracks soon appeared in many places. There was no Defects Liability Period in the case of these houses, hence the faults were not fixed, resulting in houses of different standards within the same settlement. During the field visit in 2018, a resident who had received one of these defective houses was interviewed; she reported that during the allocation, her parents had been allocated a better house some distance away from her. She was unhappy about not being able to live close to her elderly parents, and had been further disadvantaged after her father died in 2015, when the need to spend more time with her mother meant that she had to travel the distance between their houses more frequently. Additionally, she was not satisfied with the construction quality of her house, and felt that she was a loser on double accounts.

Upon closer examination of the road network on the island, although it was well-designed – with a hierarchy of large public thoroughfares to small neighbourhood lanes – the roads were not paved; because of this, during the rainy season they become muddy and movement becomes difficult. During the fieldwork in 2018, it was observed that footpaths and drains were being constructed (see Figure 4.5), and it can be expected that over time the roads will also be paved. However, the timeline for this is uncertain.

Long-term issues

A problematic and contentious issue in Dhuvaafaru is property ownership. Typically, in the Maldives land is owned by the government, and unless one buys it outright from the government, in order to live on the land and build a house or other form of dwelling, one is required to register the activity in the name of the household head. The registration can then be transferred to kin after death or during one's lifetime. The registration is the legal basis for occupying and using the land. In the case of Dhuvaafaru, there have been many disputes and disagreements in the allocation of houses, as discussed above in the section on beneficiary allocation. This has led to tensions between the residents and the government; as a result, the government has withheld the provision of registration to the properties in Dhuvaafaru.

In the Maldives, large extended families are common, especially in the more traditional setting of the outer islands. It should therefore be noted that in the construction of houses on Dhuvaafaru, there was an urban bias towards nuclear family-type housing. As per the government's policy, a family of up to eight members was given one three-bedroom house; up to 12 members was given two houses; and three houses if there were more than 12 members. Previously in Kandholhudoo, large extended families lived together in one house, or in close proximity – often in very cramped conditions. The property was registered in the name of the main head of the extended family, and not in the names of the individual families' heads. When the Dhuvaafaru settlement was built, the different families comprising an extended family each wanted an individual house, but this did not match the government's stipulation of allocation of houses by number of occupants; also, because these families did not have property registered in their names in Kandholhudoo, it led to further complications. A similar problem occurred in recovery projects on other islands (Lawther, 2014). The limited consideration of the extended family structure in the design and allocation of housing resulted in widespread modifications to the houses, discussed below in the section on transformations.

As a result of issues relating to housing allocation, a large number of the residents in Dhuvaafaru live on unregistered property; without registration, an identity card cannot be issued with the current address. This means that many residents have their former address in Kandholhudoo printed on their National Identity Cards. Furthermore, the government maintains that because they still hold registration for their previous properties in Kandholhudoo, they cannot be issued another registration – even though those properties were obliterated by the tsunami. One resident bemoaned that, 'We are refugees,' pointing to the fact that the citizens of the country are being denied a basic right and living in uncertainty on property belonging to the government – property that can be seized anytime. However, this seems unlikely to happen, given that they have been living there for a decade and will possibly be able to continue doing so over the long term.

The registration issue has also been divisive for the community, with unpleasantness between the different extended kin due to the differing numbers of houses they were able to obtain. A resident narrated that, previously in Kandholhudoo, the community was more cohesive despite the difficult living conditions, but the move to Dhuvaafaru has somewhat fragmented the community. It has become difficult to mobilise the participation of residents for a common cause, and they are inclined to relate only to those with whom they are related by kin. This has long-term implications for the community and the relationships within it.

A key issue resulting from changes in the community is migration, particularly of young persons, to Malé and other cities and tourist resorts for better employment and education opportunities. Over time, this might lead to depopulation of the island and the lack of a younger work force in future. The implications for the long-term demographic profile of Dhuvaafaru remain uncertain, and it is unclear who will benefit from all the newly built facilities on the island, who will maintain and sustain them, and who will lead new and useful initiatives.

Environmental issues

Dhuvaafaru has several options for generating renewable energy. The warm climate provides bright sunshine on many days of the year, and the coastal location provides suitable conditions for solar, wind and hydro power. IFRC had a plan to apply renewable technology provisions in Dhuvaafaru and undertook detailed feasibility studies for that purpose. However, the plan was not entirely supported by MoEE because of the limited experience in the Maldives in this sector; only photovoltaic solar power panels were eventually installed on the power station (IFRC, 2010). Nonetheless, MoEE subsequently initiated an *Outer Islands for Sustainable Energy Development Project* supported by the Asian Development Bank for 138 islands, including Dhuvaafaru (Ministry of Environment and Energy, 2014). How this project is implemented over the future will provide an indication of the prospects for renewable energy for the Dhuvaafaru community.

As discussed above, in the human-induced hazards subsection of the disaster profile, solid waste management is a serious problem in the Maldives. Garbage is indiscriminately disposed of everywhere; it seems that waste collection vessels often dump waste directly into the sea (Saleem and Sattar, 2009). There is an incessant flow of rubbish washing up on the pristine beaches of Dhuvaafaru, and there seems to be no clean-up initiative from the community (see Figure 4.6). Unless serious steps are taken to remedy this problem, it might potentially assume significant proportions in the future.

Transformations

Despite the more spacious living quarters of the new houses on Dhuvaafaru, the problem of lack of space has been intensifying with population growth; the 2014 census counted 3,016 residents in Dhuvaafaru (Maldives Bureau of Statistics, 2014), but it was roughly estimated to have exceeded 4,000 in 2018, resulting in a high density of more than 8,000 persons per square kilometre. There is

Figure 4.6 Incessant flow of solid waste onto the Dhuvaafaru coast

very little scope for expansion on this small island with demarcated zones, other than building closer to the coast – and risking exposure to coastal hazards – or by clearing vegetation and damaging the already fragile ecological balance, therefore ushering in further coastal erosion. As a consequence of these limitations, many households had extended their houses both horizontally and vertically; transformations to the original housing were widely evident throughout the island. Such transformations will, over time, place pressure on the facilities and services on the island; this is not a sustainable situation over the long run.

The houses in Dhuvaafaru built by IFRC were of good quality, with earthquake-resistant features (Oughton, 2009). However, the house design was a standard one produced by the government, applied in all the recovery projects on different islands. There was no community consultation for preparing the design, or any attempt to cater to specific household needs. The same design was used for all the houses; such a one-size-fits-all approach is problematic in catering to the diverse needs of various households of different sizes and composition, and at different stages in the family life cycle. The houses thus required adaptation and modification by households – often extensive; this was also reported in other similar projects in the Maldives (Suppasri et al., 2015). This phenomenon extended beyond the Maldives – in some of the other tsunami-affected countries similar outcomes in response to the one-size-fits-all approach were observed (O'Brien and Ahmed, 2011;

Shaw and Ahmed, 2010). Nonetheless, there was some initial effort by IFRC to obtain feedback on the design and some minor modifications were made based on the feedback; for example, an option was given to have a separate kitchen or an attached one and the community opted for an attached one with a door to the outside (IFRC, 2010), possibly for better access to home-based outdoor fish processing.

Typically, the three-bedroom 86 square metre house design was meant for small or nuclear families. However, in the Maldives large extended families are common, resulting in space problems, as narrated by Dhuvaafaru residents; this has also been reported in similar projects on other islands (Lawther, 2014). Also, as families grow over time, the need for more space arises. For example, a Dhuvaafaru resident narrated that when his family moved to the house in 2008, there were four household members including his wife and two children. Ten years later he had four more children, and with his eldest son growing and soon to start his own family, the need for a larger house became obvious. Thus, he enlarged the house, extending it up to the boundary by adding two more rooms and making the living room and kitchen larger. Because some households were extending their houses beyond the boundary, he made it a point to mention that he had confined his extension to within the boundary.

Many households have home-based livelihoods: running a shop, trades workshops, etc. However, there is very little space outside the house for this – most houses hardly have a backyard. This was also observed by others (Lawther, 2014; Zahid, 2009). Home-based fish processing – involving drying and smoking fish outdoors for personal consumption and sale – is an important cultural activity, yet there is hardly any space for this, especially after houses have been extended to accommodate growing families. As seen in recovery projects on other islands (Lawther, 2014), the community had requested larger plot sizes of about 280 square metres with such livelihoods in mind, but the government did not take this into account; instead, the standard 185.4 square metre plot size was stipulated for all of the recovery projects on the different islands.

When the community lived in Kandholhudoo, conditions inside the small houses were cramped, yet the houses had been built in such a way so that backyard space was maintained for fish processing. It appears that adequate livelihoods assessment was not conducted prior to the project on Dhuvaafaru (O'Donnell and Kakande, 2007), and this highly important cultural aspect was not taken into consideration in the project planning. Government representatives argued that a separate communal fish processing zone had been provided in Dhuvaafaru to avoid 'messiness' around the houses, but this was not a culturally acceptable solution; in practice, no one from the community used that area for fish processing. Eventually, overseas commercial firms established themselves there, processing fish and sea cucumbers for export. Sea cucumbers are not eaten in the Maldives – they are primarily an export commodity, hence such activities contribute little to the local economy of the island. In the meantime, households have adapted by making space for fish processing next to their house, sometimes making small furnaces outside the boundary along the street. Other similar income-generating transformations are extensive –

Figure 4.7 A café added to a house in Dhuvaafaru; the house on the left was extended by building right up to the boundary and adding an extra floor

houses have been converted or extended for running shops, restaurants or cafes, and trades workshops (see Figure 4.7).

On the other hand, under-occupation of homes is also evident. With many people leaving Dhuvaafaru for employment in Malé, or other urban centres and tourist resorts, some of the houses have very few residents, and sometimes even remain vacant. Additionally, some of the houses built by the government were not taken up by the beneficiaries (see section above on construction drawbacks); by the time these were built, some of the beneficiaries had established themselves elsewhere. Hence these unoccupied houses were converted to generate rental income; this was evident from the number of houses being rented out, sometimes to people from other islands who had come to Dhuvaafaru for employment, or alternatively being run as guesthouses (see Figure 4.8). Indeed, during fieldwork, accommodation was provided in one of these houses. There was a previous plan to build a guesthouse in Dhuvaafaru at a later stage, but observing this phenomenon, IFRC decided not to implement it (IFRC, 2010). This is a transformation of another nature where the original intended occupancy of the house as family residence has been changed to an income-generating asset; with continuing migration over time, this might become the future outcome for a significant proportion of the housing stock in Dhuvaafaru.

Figure 4.8 A house in Dhuvaafaru advertised for rent

There were other design problems leading to transformations. Privacy and security is important in the Maldives, and boundary walls around houses were common. Yet the houses in Dhuvaafaru were not provided with boundary walls – perhaps with the expectation that beneficiary households would build them on their own at a later stage. However, this became problematic: without any supervision or monitoring, households started building boundary walls as far as possible, sometimes even outside their plot limits, extending all the way to the street and sides, and beyond. This meant that easements were being built upon and encroaching on public land, sometimes resulting in conflict with neighbours. In general, no building regulations were being followed, and extensions and boundary walls were being built without seeking approval from the Island Council. Typically, Island Councils in the Maldives are strict and permissions are required for any construction, however small (Saleem and Sattar, 2009), and these permits can only be issued for registered land. In Dhuvaafaru, because of the registration limbo mentioned previously, the council did not interfere with the rampant transformations taking place. In fact, even some of the councillors themselves lived in unregistered houses. The situation was described by a resident: 'There is no problem with the council, no permit is taken for extensions, no one is looking. Everyone is on their own.' Indeed, on the part of IFRC such transformations were viewed in a positive light: a senior

Red Cross staff representative 'encouraged members of the community to continue improving their new homes' (Mghendi, 2009) (also see Oughton, 2009; Talbot, 2012). The extensive tree planting by residents was especially commended; despite efforts by IFRC for minimal damage to the vegetation on the island, the construction companies originally removed the vegetation extensively, so the renewal of vegetation was welcomed by IFRC (Talbot, 2012).

Perhaps this lack of monitoring or fear of reprisal from the authorities provided a sense of tenure security, encouraging the Dhuvaafaru residents to invest their personal funds to improve their property for the benefit of their families and livelihoods, and also for status and fulfilling their aspirations – such investments were often a sign of prosperity. As one resident mentioned, the houses they received provided them an opportunity to improve their economic lot, particularly in the houses converted for running small enterprises. However, not all residents had the financial resources to make such an investment. Thus, a sense of disparity also grew within the community. This was in contrast to a comment made by a resident at the early stage after just moving to Dhuvaafaru, 'there is no poor and rich – all are the same, as everyone received the same type of house' (Oughton, 2009). At that stage, the extent and nature of the transformations that would follow were not anticipated by the community or implementing agencies.

Conclusion

Unlike some of the other countries affected by the 2004 Indian Ocean Tsunami, the Maldives was not battered by gigantic waves, but inundated by an onrush of water flowing over the small, dispersed and low-lying atolls comprising the country. Although damage to property and the economy was immense, the Maldives received comparatively less attention from international media and humanitarian organisations, perhaps because the impact was less sensational. Thus, unlike the multitude of agencies that descended on other countries to provide support for relief and recovery from the disaster, in the Maldives it was primarily IFRC and its Red Cross/Crescent national society partners, particularly the American, Australian and British Red Cross, that undertook the bulk of the recovery work – such as the Dhuvaafaru project discussed in this chapter.

The Maldives faces a set of significant challenges due to its geographically scattered islands: high-density urbanisation, transport, waste management, and the need to import basic commodities and workforce. Perhaps its greatest challenge is the ongoing and impending impacts of climate change that might compound the other challenges and see a significant part of the country go under water. At the institutional level climate change-induced sea level rise is a key concern, drawing away attention from the disaster risk posed by other hazards such as earthquakes, cyclones and drought. Prior to the tsunami, and even currently to some extent, the Maldives was perceived as a low-hazard country; even though the nexus between climate change and disasters is acknowledged at the national level, there is low budgetary allocation for disaster resilience. Thus, NDMC – the nationally mandated governmental body responsible for dealing with disasters – is faced with the task of seeking international support to advance the disaster resilience agenda.

The outcome of the above scenario is manifested in the post-tsunami recovery project in Dhuvaafaru, a partnership between NDMC and IFRC. This project evinces many positive aspects, a 'build back better' initiative that has reduced the disaster risk of a community that previously lived on a precarious island settlement that was decimated by the tsunami. A range of facilities and housing on the newly built island of Dhuvaafaru was praised by both agencies and the beneficiary community. However, a new set of predicaments have emerged, particularly relating to the uncertain status of property ownership, within which the self-initiatives of residents to transform their built environment brings a new dimension to the recovery process. Despite the disaster resilience efforts incorporated into the project, which earned it the reputation of being a 'safe island', the protection offered to this far-flung island in the middle of the Indian Ocean is limited in the face of the imminent threat posed by climate change. Together with the other inhabitants of the Maldives, the Dhuvaafaru residents will need a concerted effort to stay afloat.

References

Afzal, M.S., Solandt, J., Zareer, I.H. and Hammer, M. (2016) *Little and Large: Surveying and Safeguarding Coral Reefs & Whale Sharks in the Maldives*. Melbourne: Biosphere Expeditions.

Ahmed, I. (2017) 'Flood Adaptations in the Asian Vernacular' in Piesik, S. (ed.) *Habitat: Vernacular Architecture for a Changing Planet*. London: Thames & Hudson, pp. 508–513.

Ahmed, I. and McEvoy, D. (2014) 'Post-Tsunami Resettlement in Sri Lanka and India: Site planning, Infrastructure and Services'. *Journal of Disaster Resilience in the Built Environment*, 1(5), pp. 53–65.

Arup International Development. (2012) *Dhuvaafaru Island Tsunami Resettlement Study*. Available at: www.arup.com/projects/dhuvaafaru-island-tsunami-resettlement-study (accessed 25 June 2019).

Asian Development Bank (ADB). (2005) *From Disaster to Reconstruction: A Report on ADB's Response to the Asian Tsunami*. Manila: ADB.

Bluepeace. (2009) *Dhuvaafaru: One of the Most Vulnerable Islands to Climate Change*. Available at: www.bluepeacemaldives.org/blog/climate-change/dhuvaafaru-vulnerable-to-climate-change (accessed 3 July 2019).

Burke, J. and Rasheed, Z. (2015) *Maldives Threatens Migrant Workers with Expulsion over Planned Protest*. The Guardian. Available at: www.theguardian.com/world/2015/mar/26/maldives-migrant-workers-protest (accessed 27 July 2019).

Cheung, W., Watson, R. and Pauly, D. (2013) *Signature of Ocean Warming in Global Fisheries Catch*. Available at: www.nature.com/articles/nature12156 (accessed 1 July 2019).

countryeconomy.com. (2004) *Maldives – Population 2004*. Available at: https://countryeconomy.com/demography/population/maldivas?year=2004 (accessed 30 June 2019).

Faisal, M. (2008) *Living in a Crowded Island: Urban Transformation in the Maldives* (unpublished research proposal). Wellington: Victoria University of Wellington.

Fernandez, G. and Ahmed, I. (2019) '"Build Back Better" Approach to Disaster Recovery: Research Trends since 2006'. *Progress in Disaster Science*, 1, pp. 1–4.

Fritz, H.M., Borrero, J.C., Synolakis, C.E. and Yoo, J. (2006) '2004 Indian Ocean Tsunami Flow Velocity Measurements from Survivor Videos'. *Geophysical Research Letters*, 33(24), pp. 1–5.

Hamid, I. (2015) *Permanent Water Crisis for Maldives Islanders*. ReliefWeb. Available at: https://reliefweb.int/report/maldives/permanent-water-crisis-maldives-islanders (accessed 2 July 2019).

Hay, J.E. (2006) *Climate Risk Profile for the Maldives*. Malé: Republic of the Maldives.

Hello Publishing. (2018) *Hello Maldives*. Singapore: Hello Publishing.

Intergovernmental Panel on Climate Change (IPCC). (2012) *Managing the Risks of Extreme Events and Disasters to Advance Climate Change Adaptation*. Cambridge: Cambridge University Press.

International Federation of Red Cross and Red Crescent Societies (IFRC). (2008) *South/Southeast Asia & East Africa: Earthquake and Tsunamis*. Geneva: IFRC.

International Federation of Red Cross and Red Crescent Societies (IFRC). (2010) *Dhuvaafaru Safe Island Project - Evaluation Phase I*. Geneva: IFRC.

International Federation of Red Cross and Red Crescent Societies (IFRC). (n.d.) *Maldives: Rebirth of a Community*. Available at: www.ifrc.org/en/what-we-do/disaster-management/responding/ongoing-operations/indian-ocean-tsunami3/affected-countries/maldives-rebirth-of-a-community/ (accessed 26 July 2019).

International Federation of Red Cross and Red Crescent Societies (IFRC) (n.d.) *Tsunami Two-Year Progress Report: Maldives*. Geneva: IFRC.

Lawther, P.M. (2010) 'Community Involvement in Post Disaster Re-construction – Case Study of the British Red Cross Maldives Recovery Program'. *International Journal of Strategic Property Management*, 13(2), pp. 153–169.

Lawther, P.M. (2014) 'Permanent Housing in Community Socio-Ecological Recovery: The Case of T. Vilufushi, Maldives'. *Open House International*, 39(3), pp. 28–36.

Majeed, A. (2018) *The Migrant Workers of the Maldives*. Available at: https://maldivesindependent.com/in-pictures/the-migrant-workers-of-the-maldives-135867 (accessed 30 June 2019).

Maldives Bureau of Statistics (MBS). (2014) *Population and Housing Census*. Malé: MBS.

Maldives Independent. (2016) *Light Earthquake Strikes 300km East of Maldives*. Available at: https://maldivesindependent.com/environment/light-earthquake-strikes-300km-east-of-maldives-127116 (accessed 30 June 2019).

Maldives Independent. (2017) *Damage to 60 Islands After Extreme Weather*. Available at: https://maldivesindependent.com/environment/damage-to-60-islands-after-extreme-weather-134409 (accessed 30 June 2019).

Maldives Times. (2019) *The Inventory of Accidents of the Maldives Times*. Available at: https://maldivestimes.com/category/disasters/accidents/page/6/(accessed 3 July 2019).

Mangroves for the Future. (n.d.) *Maldives*. Available at: www.mangrovesforthefuture.org/countries/members/maldives/(accessed 3 July 2019).

Marti, R.Z. (2005) *The 2004 Hurricanes in the Caribbean and the Tsunami in the Indian Ocean*. New York: United Nations.

Mghendi, N. (2009) *Maldives: Dhuvaafaru Celebrates Its 'Birthday'*. Available at: www.ifrc.org/en/news-and-media/news-stories/asia-pacific/maldives/maldives-dhuvaafaru-celebrates-its-birthday/(accessed 18 September 2019).

Ministry of Construction and Public Infrastructure. (2008) *Maldives National Building Code Handbook*, 1st ed. Malé: Ministry of Construction and Public Infrastructure.

Ministry of Environment and Energy (MoEE). (2014) *Republic of Maldives: preparing Outer Islands for Sustainable Energy Development*. Malé: MoEE.

Ministry of Fisheries and Agriculture. (n.d.) *Atolls of Maldives*. Available at: www.atollsofmaldives.gov.mv/ (accessed 3 July 2019).

Ministry of Planning and National Development (MoPND). (2005) *The Maldives One Year after the Tsunami*. Malé: MoPND.

Mohammed, M. (2015) *Migrant Workers' Voice: Illegal and Silenced in the Maldives*. AP Migration. Available at: http://apmigration.ilo.org/news/migrant-workers2019-voice-illegal-and-silenced-in-the-maldives (accessed 25 July 2019).

Moosa, F. (2014) *Country Report: Republic of Maldives*. Kobe: Asian Disaster Reduction Centre (ADRC).

National Bureau of Statistics (NBS). (2006) *Analytical Report: Census 2006*. Malé: NBS.

National Bureau of Statistics (NBS). (2014) *Total Maldivian Population by Islands*. Malé: NBS.

National Bureau of Statistics (NBS). (2018) *Statistical Yearbook of Maldives 2018*. Malé: NBS.

National Disaster Management Centre (MDMC). (2010) *Strategic National Action Plan for Disaster Risk Reduction and Climate Change Adaptation 2010-2020*. Malé: NDMC.

National Disaster Management Centre (NDMC). (2014) *Cases Reported in 2014*. Malé: NDMC.

National Disaster Management Centre (NDMC). (2016) *CBDRM Programs*. Available at: http://ndmc.gov.mv/projects/new-page-2/ (accessed 18 September 2019).

Naylor, A.K. (2015) *Island Morphology, Reef Resources, and Development Paths in the Maldives*. SAGE Journals. Available at: https://journals.sagepub.com/doi/abs/10.1177/0309133315598269 (accessed 16 July 2019).

New World Encyclopedia. (2018). *Maldives*. Available at: www.newworldencyclopedia.org/entry/Maldives (accessed 16 Jul. 2019).

O'Brien, D. and Ahmed, I. (2011) 'Donor-Driven Housing, Owner-Driven Needs' in Kaufman, N. (ed.) *Pressures and Distortions: City Dwellers as Builders and Critics: Four Views*. New York: Raphael Vinoly Architects PC, pp. 237–337.

O'Donnell, I. and Kakande, C. (2007) *Learning from the Tsunami: Mid-Term Review - Findings Overview*. Geneva: International Federation of Red Cross and Red Crescent Societies (IFRC).

Oaku, Y. (2018) *Overcoming Trauma: Safety Drills in the Maldives Help Tsunami Survivors to Make Peace with their Past*. Medium. Available at: https://medium.com/@UNDPasiapac/overcoming-trauma-safety-drills-in-the-maldives-help-tsunami-survivors-to-make-peace-with-their-652a5b4c3bea (accessed 3 July 2019).

Oughton, S. (2009) *Maldives Tsunami Survivors Move back to 'Beloved Island'*. Available at: www.ifrc.org/en/news-and-media/news-stories/asia-pacific/maldives/maldives-tsunami-survivors-move-back-to-beloved-island/ (accessed 7 February 2020).

PreventionWeb. (2014). *Maldives - Disaster & Risk Profile*. Available at: www.preventionweb.net/countries/mdv/data/(accessed 26 June 2019).

Risk Management Solutions India (RMSI). (2006) *Developing a Disaster Risk Profile for Maldives*. Malé: United Nations Development Programme (UNDP) Maldives.

Riyaz, M. and Park, K.H. (2010) '"Safer Island Concept" Developed after the 2004 Indian Ocean Tsunami: A Case Study of Maldives'. *Journal of Earthquake and Tsunami*, 4(2), pp. 135–143.

Saleem, M. and Sattar, S.A. (2009) *Study on Post-Tsunami Restoration and Conservation Projects in the Maldives*. Malé: International Union for Conservation of Nature (IUCN) Maldives Project Office.

Shaw, J. and Ahmed, I. (2010) *Design and Delivery of Post-Disaster Housing Resettlement Programs: Case Studies from Sri Lanka and India*. Canberra: Australian Research Council.

Shaw, R. (ed.) (2014) *Recovery from the Indian Ocean Tsunami: A Ten-Year Journey*. Tokyo: Springer.

Sphere Project. (2004) *Humanitarian Charter and Minimum Standards in Disaster Response*. Geneva: Sphere Project.

Storlazzi, C.D., Gingerich, S.B., Dongeren, A.V., Cheriton, O.M., Swarzenski, P.W., Quataert, E., Voss, C.I., Field, D.W., Annamalai, H., Piniak, G.A. and McCall, R. (2018). *Most Atolls will be Uninhabitable by the Mid-21st Century because of Sea-Level Rise Exacerbating Wave-Driven Flooding*. Science Advances. Available at: https://advances.sciencemag.org/content/4/4/eaap9741 (accessed 25 July 2019).

Suppasri, A., Goto, K., Muhari, A., Ranasinghe, P., Riyaz, M., Affan, M., Mas, E., Yasuda, M. and Imamura, F. (2015) 'A Decade after the Indian Ocean Tsunami: The Progress in Disaster Preparedness and Future Challenges in Indonesia, Sri Lanka, Thailand and the Maldives'. *Pure and Applied Geophysics*, 172, pp. 3313–3341.

Talbot, J. (2012) *Returning to Dhuvaafaru, Eight Years on from the Indian Ocean Tsunami*. Available at: www.ifrc.org/en/news-and-media/news-stories/asia-pacific/maldives/returning-to-dhuvaafaru-eight-years-on-from-the-indian-ocean-tsunami-60490/ (accessed 7 February 2020).

TripAdvisor. (2012) *Earthquake Felt in Malé – Maldives Forum*. Available at: www.tripadvisor.com.au/ShowTopic-g293953-i7445-k5320387-o10-Earthquake_felt_in_Male-Maldives.html (accessed 16 July 2019).

United Nations Development Programme (UNDP). (2006) *Developing a Disaster Risk Profile for Maldives*, 1st ed. Malé: UNDP Maldives.

United Nations Development Programme (UNDP). (2008) *Detailed Island Risk Assessment in Maldives: Natural Hazard and Physical Vulnerability Assessment Report*. Malé: UNDP Maldives.

United Nations Environment Programme (UNEP). (2005) *Maldives Post-Tsunami Environmental Assessment*. Nairobi: UNEP.

United Nations Framework Convention on Climate Change (UNFCCC). (2001) *First National Communication of the Republic of Maldives to the United Nations Framework Convention on Climate Change*. Malé: Ministry of Home Affairs, Housing and Environment.

vNews. (2016) 25-Storey IGMH Building Will Be the Tallest Establishment in the Country: President. Available at: www.vnews.mv/64951 (accessed 17 September 2019).

Wadey, M., Brown, S., Nicholls, R. and Haigh, I. (2017) *Coastal Flooding in the Maldives: An Assessment of Historic Events and Their Implications*. ePrints Soton. Available at: https://eprints.soton.ac.uk/412275/ (accessed 25 July 2019).

Wikipedia. (2018) *Dhuvaafaru*. Available at: https://en.wikipedia.org/wiki/Dhuvaafaru (accessed 26 July 2019).

World Bank (WB). (2018) *Maldives Overview*. Available at: www.worldbank.org/en/country/maldives/overview (accessed 30 June 2019).

World Bank (WB), Asian Development Bank (ADB) and UN System (UNS). (2005) *Tsunami: Impact and Recovery (Joint Needs Assessment)*. Washington, DC, Manila and New York: WB, ADB and UNS.

World Health Organization. (2012) *Tropical Cyclone Nilam: Maldives, Sri Lanka, India*. Available at: www.searo.who.int/entity/emergencies/crises/nilam/en/ (accessed 18 September 2019).

World Population Review. (2019) *Maldives Population 2019*. Available at: http://worldpopulationreview.com/countries/maldives-population/(accessed 24 September 2019).

Zahid, A. (2009) *Hussain Alifulhu: A Brighter Future in Dhuvaafaru, Maldives*. Available at: www.ifrc.org/en/news-and-media/news-stories/asia-pacific/maldives/hussain-alifulhu-a-brighter-future-in-dhuvaafaru-maldives/ (accessed 18 September 2019).

Zahir, M. (2011a) *Population Explosion: A Major Environmental Issue in Malé*. ECOCARE Maldives. Available at: http://ecocare.mv/population-explosion-a-major-environmental-issue-in-male%E2%80%99/ (accessed 30 June 2019).

Zahir, M. (2011b) *Beach Erosion: A Vulnerable Scenario in the Maldives*. ECOCARE Maldives. Available at: http://ecocare.mv/beach-erosion-a-vulnerable-scenario-in-the-maldives/(accessed 2 July 2019).

2015 Ghorka Earthquake

Building codes for resilience to earthquakes in Nepal

5 Rule of thumb

Opportunities and barriers to compliance of building codes for disaster resilience in Nepal

Introduction

Nepal is affected by many natural hazards, including disastrous earthquakes – the most recent of which was the April 2015 earthquake or Gorkha Earthquake. Nepal Disaster Report (2017) suggests that, every year in the last 45 years, the incidences of disasters have been growing in Nepal; this has had a profound negative impact on people, economy and the environment. Analysis of available data also shows that disasters annually diminish about 2 per cent of national GDP. Within the South Asia region, Nepal has the largest mortality rate due to disasters. According to the Ministry of Home Affairs (2018, p. xii), 'More than 80 per cent of the total population of Nepal is at risk from natural hazards, such as floods, landslides, windstorms, hailstorms, fires, earthquakes and Glacial Lake Outburst Floods.'

This chapter explores the implementation of safe building codes to enhance disaster resilience in Nepal, drawing from the recovery and reconstruction activities undertaken after the 2015 earthquake. A number of initiatives have been undertaken to make improvements at the institutional level, together with community-driven interventions, but these have had mixed results. Despite a flexible 'rule of thumb' approach, critical barriers to implementation of building codes persist. Recovery from the 2015 earthquakes was faced with many unique sets of political complexities and challenges due to socio-cultural and economic conditions. The post-disaster initiatives demonstrate a complex, but consistently coherent, theme relating to post-disaster recovery in South Asia.

This chapter contains two main components: (a) an overview of the demographics of Nepal followed by a disaster risk profile; and (b) an analysis of the impacts of the earthquake and policy framework for disaster resilience in relation to a case study of implementation of building codes to improve earthquake-resilient infrastructure. The first component delineates the opportunities and challenges for achieving disaster resilience in the country; the second component, the case study, provides an understanding of how disaster resilience initiatives are applied in practice via a recovery project.

The case study is based on the 2015 Gorkha Earthquake, underpinned by the analysis of secondary data from published materials and primary data collected through fieldwork and interviews. Published materials included academic papers, industry/government reports and newspaper articles. The main primary data collection method included stakeholder and key informant interviews, and photographic and audio-visual recordings. The interviews were undertaken with people who had experience in earthquake reconstruction, building codes and planning in Nepal. The key interviewees were engineers and government officials working in the field of building code implementation and building construction, and politicians who play an important role in urban development and housing. Interview data was analysed to identify emergent themes.

Disaster overview: 2015 Gorkha Earthquake

On 25 April 2015, a magnitude 7.8 earthquake struck Nepal, just east of Gorkha district. The earthquake was Nepal's worst in 81 years, causing the deaths of about 9,000 people, injuring 18,000 and displacing more than 500,000; 1,000 health facilities were also damaged (Mishra and Adhikari, 2019; Rafferty, 2015). The Gorkha Earthquake destroyed more than 500,000 houses and damaged another 250,000 houses (NPC, 2015). It also caused landslides and avalanches, crippled transportation networks, levelled villages close to the epicentre, ruined numerous heritage sites and destroyed buildings across Nepal. The strongest impact was felt in remote rural areas, making the response extremely challenging. Over the years, recurrent earthquakes in Nepal have resulted in structural vulnerability of buildings, making them susceptible to collapse during earthquakes (Gautam et al., 2016). Most of the masonry buildings that collapsed during the 2015 earthquake were probably the buildings that survived the past earthquakes of 1934 and 1988 (Gautam et al., 2016) (see Figure 5.2). The National Planning Commission (2015) estimated that the total earthquake damage to existing stock of assets was almost US$5 billion. In addition, an estimated US$2 billion of economic losses resulted due to destruction of assets. With both figures combined, this represented an economic impact or loss equivalent to about one third of Nepal's GDP and well over 100 per cent of the Gross Fixed Capital Formation (NPC, 2015).

The Gorkha Earthquake was extensively covered by the media and has been widely represented in the literature. Some of the main events that followed the earthquake were highlighted by Reid (2018) and Rafferty (2015) and are summarised here: within a day of the earthquake, international emergency response teams arrived. Soon after, a District Disaster Relief Committee (DDRC) was established to coordinate distribution of relief supplies through a 'one door' policy. The earthquake was followed by several aftershocks, notably on 12 May 2015, when a magnitude 7.3 aftershock occurred (see Figure 5.1), causing further massive destruction to critical and social infrastructure and residential properties. There were 31 districts affected, of which 14 were declared 'crisis-hit'. A post-disaster needs assessment was undertaken (He, 2019; NPC, 2015), which helped facilitate a pledge by international donors for about US$4.4 billion. The monsoon season during June to September exacerbated the crisis situation by causing landslides and creating inaccessibility in the

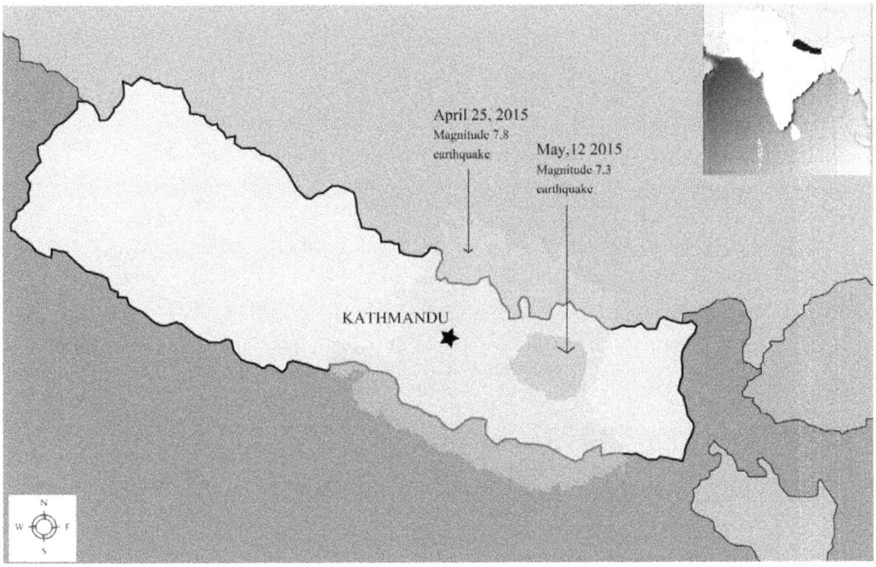

Figure 5.1 Map of Nepal showing the epicentre of the Gorkha Earthquake and the subsequent aftershock

mountainous terrain, with thick clouds impeding aerial operations. In the midst of this turmoil, during September 2015 a new national constitution was approved after a decade of delays, leading to a restive atmosphere with many political demonstrations; for nearly five months after that, land travel between Nepal and India was restricted, consequently resulting in shortages of essential relief supplies. At the end of 2015, the government eventually established a National Reconstruction Authority to oversee the recovery process, which was initially rejected in parliament earlier in the year because of political pressures.

Country overview of key issues

Nepal is located in a geographically challenging area lying in the central part of the Himalayas. Bordered by China and India, Nepal is home to eight of the world's highest mountains including Mount Everest, known to Nepali people as 'Sagarmatha'. Although the overall population density is relatively low across Nepal compared to some of the other South Asian countries in this book, the population is mostly concentrated in the southern-most plains of the Terai region and the central hilly region. Nepal is classified as a low-income country with a population close to 30 million (World Bank, 2019; World Population Review, 2019). Nepal is comprised of seven provinces administered by the federal government. However, the sub-level administration is more complex, with six metropolitan cities, 11 sub-metropolitan cities and 276 municipalities,

Figure 5.2 A building in Kathmandu damaged by the 2015 Earthquake; in 2017 it was found to be occupied because of the slow reconstruction process

together forming 293 urban municipalities and 460 rural municipalities (Quora, 2018). Until 2014, Nepal had only 58 municipalities; the increase in municipalities reflects the urban population growth (see Figure 5.3) and consequent increase in administrative governance.

Political backdrop

Nepal is a federal democratic republic with an elected President as head of state; the executive consists of a Prime Minister and Cabinet chosen by the Prime Minister in consultation with the governing party and any coalition partners (The Economist, 2017). However, Nepal remained mostly closed to the outside world until 1950. Nepal was admitted to the United Nations in 1955. In 1991 the kingdom established a multi-party parliamentary system. In 2008, however, after a decade-long period of violence and turbulent negotiation with a strong Maoist insurgency, the monarchy was dissolved and Nepal was declared a democratic republic (Rose et al., 2019).

According to Deraniyagala (2005, p. 47) the 'conflict [inherently] is the outcome of an uneven process of development that led to the social and economic exclusion of large segments of the population.' Sharma (2006, p. 1237) argues

Figure 5.3 Urban growth of the Kathmandu Valley has necessitated expansion of municipalities and urban governance

further that 'development failure, together with corruption and short-sightedness of the ruling elite, has provided fertile grounds for the civil war in Nepal.' The development efforts in Nepal that did not reach the poor or address the rise in unemployment, poverty and rural–urban inequality was the driver for the Maoist conflict/insurgency from 1996 to 2006, as the Maoists aimed to establish a People's Republic (Bohara et al., 2006). The ten-year long armed conflict between the Communist Party of Nepal (Maoist) against the Government of Nepal resulted in the deaths of many people, including civilians, insurgents, army and police personnel; the conflict also displaced hundreds of thousands of people, mostly in rural Nepal. The conflict ended with the signing of the Comprehensive Peace Accord in 2006. It has been suggested that the conflict could have been avoided if the political leaders and ruling elite had foreseen the consequences of socio-economic disparity and exclusion (Sharma, 2006). Underlying tensions continue to persist regarding inclusion and the lack thereof (Thapa and Ramsbotham, 2017).

During the Maoist conflict, the assassination of King Birendra on 1 June 2001 created more instability in Nepal. Twelve members of the family were killed in a mass shooting during a monthly reunion dinner of the royal family. The shooter was another family member, King Birendra's son Crown Prince Dipendra. At the end of the shooting rampage Dipendra shot himself, but did not die – instead, he slipped into coma, and as the king had been killed, Dipendra was declared King of Nepal. He died in hospital three days later without regaining consciousness. Birendra's brother, Gyanendra, became king again after the massacre and the death of his nephew King Dipendra. The circumstances of the massacre remain controversial, even today. With more turmoil in 2008 the 340-year rule of the monarchy was abolished. Until 2007, when the interim constitution was approved by the parliament, Nepal was a constitutional Hindu monarchy. However, the current constitution, ratified in September 2015 by the parliament, indicates the country as a secular democratic republic. The assignation along with the insurgency and the political instability have further impacted Nepal's development.

Economy

Nepal's economy relies heavily on foreign aid and tourism. Nepal's GDP per capita (2018) is US$1026 and this, according to World Bank statistics, is just above the lowest income countries that have a per capita GDP of less than US$900 (World Bank, 2019). The Index of Economic Freedom (2019) indicates that Nepal's long-term development and foreign investment, despite a few free trade agreements, is held back by some fundamental structural issues. The lack of entrepreneurial spirit, political uncertainty, a history of state intervention and a difficult business climate have contributed to long-term developmental issues. Moreover, the onerous approval process, a lack of transparency, and property rights undermined by an inefficient judicial system that is subject to undue and corrupt influence have also contributed to poor development.

The International Monitory Fund (IMF) (2018) indicated that Nepal's economic growth was expected to reach 6.5 per cent in that fiscal year, supported by greater political stability. This growth percentage is in the increasing trajectory, as the average growth was around 4.5 per cent since 2006. IMF suggested that this expansion also comes with some challenges and the need for measured tightening of policies to safeguard macroeconomic and financial stability. A report by Oxfam International and Humanitarian Accountability Monitoring Initiative (2019) provides a challenging economic picture for Nepal in terms of income inequality and gender inequality, impacting on development and disaster recovery. The report suggests that although the poverty levels have significantly reduced in Nepal, the inequality of wealth is substantial: the richest 20 per cent hold 56.2 per cent of all wealth, while the poorest 20 per cent share 4.1 per cent of wealth (Oxfam International and HAMI, 2019).

Nepal is a largely rural country, with 83 per cent of the population living in rural areas (Jones et al., 2016). The country rapidly urbanised between 1990 and 2014, with an urbanisation growth rate of 3 per cent during this period (Shrestha et al., 2018). Nepal ranks 136th on the Heritage Foundation's 2019 economic freedom index (Heritage Foundation, 2019). The Human Development Index, on the other hand, ranks Nepal 145 out of 185 countries (Jones et al., 2016; Regmi, 2016). The country's population is estimated to be 29.3 million people (Heritage Foundation,

2019). The annual GDP growth rate has seen an increase from 3.5 per cent in 2009 (Dangal, 2011) to 4.3 per cent in 2017 (Heritage Foundation, 2019). Nepal has a somewhat complex geophysical structure, prone to a myriad of disasters (Gaire et al., 2015; World Economic Forum, 2015). The demographic and economic challenges heighten the disaster risk in a country that has significant exposure to natural hazards. During the past decade, the country introduced earthquake risk management initiatives that have increased awareness and built the capacity of stakeholders across the country (Dixit et al., 2018).

Disaster risk profile

Nepal ranks as one of the most disaster-prone countries in the world, in terms of relative vulnerability to climate change, earthquakes and flood-related hazards (4th, 11th and 30th out of 200, respectively) (Bhattacharjee et al., 2017; Dangal, 2011; Gaire et al., 2015; Nepal et al., 2018). Nepal's disaster risk profile is ranked 31 out of 191 countries in INFORM's risk index (Index for Risk Management, n.d.). Between 2000 and 2011, disasters in Nepal killed at least 4,000 people (Dangal, 2011). Nepal is exposed to about 500 disaster events annually (Nepal et al., 2018). Between 2005 and 2015, over 1.4 million people were injured and approximately 23 million made homeless as a result of disaster events (Nepal et al., 2018).

Nepal also faces a number of environmental problems, such as soil and water acidification, deforestation and forest degradation due to extraction of firewood. Contaminated water from improper human and animal waste management, agricultural run-off, industrial effluents and unmanaged solid waste are also contributing to elevated disaster risks.

Although the recent 2015 Gorkha Earthquake brought into focus the seismic risk that the country faces, there are also other natural hazards including severe thunderstorms, floods, landslides, drought/water scarcity, extreme heat, wildfires and famine, depending on the timing, intensity and duration of the summer monsoons (Global Facility for Disaster Risk Reduction, 2019). Some of the key hazards are discussed below.

Floods

Flooding is a regular occurrence during the monsoon season, mostly due to the overflow of rivers (Dixit, 2013). The country also experiences *bishyari* events – damage to dams caused by landslides falling directly into rivers – and glacial lake outburst floods (GLOFs), especially in the upper Himalayan region (Dixit, 2013; Matambo and Shrestha, 2013). Despite the high prevalence of floods in Nepal, they were traditionally managed on an ad-hoc basis and attended to as and when they occurred (Jones et al., 2014).

Landslides

Landslides usually occur due to heavy rainfall and earthquakes (Dangal, 2011). In July 2019, the International Federation of Red Cross and Red Crescent

Societies (IFRC) reported that heavy rainfall had triggered flooding and landslides in some districts of Nepal (IFRC, 2019). Repeatedly, the occurrence of heavy flooding is directly linked to landslides in Nepal. These events resulted in the deaths of 78 people, while 76,990 families were affected and 8,798 families were displaced. Housing was also heavily damaged: 5,922 houses were completely destroyed while 12,772 houses were partially damaged (IFRC, 2019). In 2014, Jure village of Sindhupalchok District experienced a huge landslide that killed 150 people, disrupting transportation and affecting infrastructure in the district (Gaire et al., 2015).

Climate change impacts

The Government of Nepal has reported that the country is highly vulnerable to climate change (Ministry of Population and Environment, 2016). Climate modelling exercises conducted in 2009 revealed that the mid-hills region in the country is likely to become more arid during the non-monsoon seasons (Dixit, 2013). The impact of climate change is expected to elevate temperatures while making precipitation erratic (Dixit, 2013). Annual precipitation is predicted to decrease from 10 to 20 per cent across the country (MoPE, 2016). Should the monsoon pattern in Nepal change, the country may experience failures in rural livelihood, urban food supply and water management, which the local populations rely on (Dixit, 2013). The direct impact of climate change and extreme events on Nepal is projected to be equivalent to 1.5 to 2 per cent of the national GDP per year (MoPE, 2016).

Earthquakes

Nepal is one of the most earthquake-prone regions in the world, as it lies at the boundary of the actively converging Indo-Australian plate and Eurasian Plate. Furthermore, Nepal has regularly witnessed major earthquakes as it lies in an active seismic zone between the Indian and Eurasian plates (MoHA, 2015). The first recorded event was in 1255 AD; subsequent major earthquake events were recorded at least once every hundred years from the 1200s to the 1800s; in recent history, major earthquakes occurred in 1934, 1966, 1980, 1988, 2011, as well as the recent earthquake in 2015 (Chamlagain and Gautam, 2015). One of the last great earthquakes was in 1934 – it measured over 8.0 on the Richter scale and resulted in the death of 8,519 people, with 207,740 structures damaged (Rana, 2015). The 1988 Eastern Nepal earthquake killed 721 people and injured 6,553; during this event, 20,000 buildings collapsed or were damaged, including 948 schools. The impacts from earthquake events are often devastating – casualties and damage from the history of earthquakes in the Kathmandu Valley are shown in Table 5.1.

Major earthquakes have been recorded in Nepal since 1223 AD (Subedi et al., 2016). As discussed earlier in the disaster overview section, on 25 April 2015, a 7.8 magnitude earthquake hit central Nepal and its environs (Dixit et al., 2018;

Table 5.1 Casualties and damages in past earthquakes in the Kathmandu Valley (adapted from: Ahmed et al., 2018)

Year	*Date*	*Earthquake epicentre*	*Casualties*		*Buildings/Temples*	
			Deaths	Injuries	Collapsed	Damaged
1988	21 Aug.	Udayapur	721	6,553	650	1,814
1934	15 Jan.	Bihar/Nepal	8,519	–	12,397	43,342
1837	17 Jan.	–	–	–	–	–
1833	26 Aug.	–	43	30	18,000	–
1823	–	–	–	–	–	–
1810	May	–	Moderate casualties		Many buildings and temples collapsed	
1767	June	–	–	–	–	–
1681	–	–	–	–	Many buildings and temples collapsed	
1408	–	–	Heavy casualties		Many buildings and temples collapsed	
1260	–	–	Heavy, widespread famine and epidemic		–	–
1255	7 Jun	–	Around one-third of total population killed, including King Abhaya Malla		Many buildings and temples collapsed	

Gautam et al., 2016). This was immediately followed by an aftershock of magnitude 6.7 (Gautam et al., 2016). Another strong aftershock of magnitude 7.3 occurred in central Nepal on 12 May 2015 (Gautam et al., 2016).

From the early 2000s, earthquake disaster risk in the Kathmandu Valley region has been increasing significantly, mainly due to uncontrolled development, poor construction practices and lack of awareness (Dixit et al., 2013). It was estimated at that time that as many as 60 to 75 per cent of buildings in the Kathmandu Valley and Bhaktapur were likely to be damaged heavily and some even beyond repair in the event of an earthquake (Dixit et al., 2013). It is therefore not surprising that a huge number of buildings in the Kathmandu Valley and in other districts – such as Gorkha, Bhaktapur, Lalitpur, Sindhupalchowk, Sindhuli, Ramechhap, Dolakha, Nuwakot, Dhading, Rasuwa Solukhumbu and Kavre – collapsed during the 2015 Gorkha Earthquake (IFRC, 2015; Ruszczyk, 2018).

The 2018 INFORM risk index indicates that, among all the hazards, earthquakes contribute the most to Nepal's risk profile (9.9 out of a 10 scale). In this context, safe building codes in Nepal can play a vital role in achieving disaster resilience by supporting the construction of earthquake-resistant buildings. However, most of the building infrastructures in Nepal have been constructed by households themselves, generally not licensed to build, based on their own needs and budget, and without consulting engineers; several studies have shown that many of these structures are likely to be inadequate in resisting earthquakes. Though an earthquake design code exists in Nepal, it is not mandatory in every part of the country and people do not follow it properly even where it is mandatory (Pokharel and Goldsworthy, 2017). The case study in this chapter focuses on earthquakes, given that they are covered in the National Building Code of Nepal. Other hazards such as floods, landslides and drought also affect Nepal, but are beyond the scope of the case study here.

Case study: implementation of building codes after the 2015 Gorkha Earthquake

After the devastating 2015 Gorkha Earthquake, the post-disaster events have had a significant impact on the administrative structure of Nepal and institutional approaches to disaster resilience. The key initiatives developed after the event focused on improved building codes, better urban planning and assisting with better engineered construction. However, Nepal still faces significant challenges in mitigating disaster risks at a practical level due to socio-cultural and economic issues. This case study reviews the events and experiences following the 2015 earthquake and identifies lessons that can be learned from Nepal. Some of the major problems include high-level ones, such as an unstable political situation, ongoing constitutional reform, ill-equipped institutions and socio-economic constraints, as well as operational problems such as training-skilling, improving institutional capacities and improving the governance framework for constructing safe buildings (Daly et al., 2017).

The changes to Nepal's geographic administrative units after the 2015 earthquake posed some challenges in terms of conducting an analysis based on administrative units. In 2016, 205 new municipalities in Nepal had experienced the implementation of the National Building Code (NBC) while some of the other municipalities had implemented the NBC prior to that. The interviews were conducted in both the old municipalities – those with more experience of the implementation of building codes – and new municipalities, which have only recently started the implementation. A total of five municipalities were selected for the research, including three municipalities in Kathmandu Valley: Lalitpur – the first municipality to implement NBC – and two new municipalities, Thimi and Budhanilkhanta (See Figure 5.4). The other two municipalities selected were located outside Kathmandu Valley: Dharan, which implemented the NBC in 2007, and Duhabi, a new municipality.

Post-disaster response and reconstruction

In the immediate aftermath of the 2015 earthquake, Nepali society quickly reacted to the situation, mobilising national and local businesses, as well as volunteer groups to deliver much-needed support, especially in remote communities (Naujoks, 2016; World Economic Forum, 2015). Emergency workers, security forces (for example, the Nepalese Army, Nepal Police and Armed Police Force) as well as

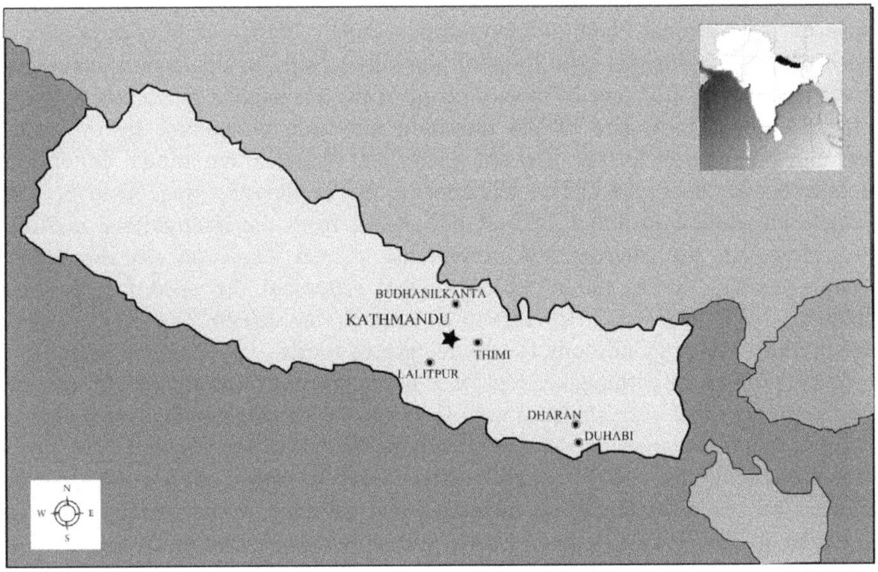

Figure 5.4 Map of Nepal showing locations relevant to this chapter

residents and bystanders immediately worked to rescue survivors from the debris and clear the rubble (IFRC, 2015). This swift response provided support for many people (Shrestha and Pathranarakul, 2018; World Economic Forum, 2015). However, there were concerns that most of the rescue operations were limited to the Kathmandu Valley and its environs, leaving neighbouring communities without water, food, tents and medical supplies (Neupane, 2015). The Government appeared unprepared to handle the scale of the disaster (Pokharel et al., 2018).

Within a few days of the earthquake, many international institutions and humanitarian organisations arrived in Nepal to provide relief to affected communities. However, with a lack of governmental direction in the immediate post-earthquake context, sub-optimal relief distribution outcomes were evident. The proposal of the government to coordinate relief distribution through a 'one-door' policy via the District Disaster Relief Committee (DDRC) was impeded by low technical capacity, political conflicts and biases (Hagelsteen and Burke, 2016; Lord and Moktan, 2017; Shrestha and Pairote, 2018). The main objectives of DDRC were to manage excessive relief and overcrowding of donors, and to enable appropriate distribution throughout affected geographical areas; however, claims of marginalisation of disaster victims based on class overshadowed the effectiveness of distribution. Preferential treatment and biases emerged due to the absence of functional systems to organise and distribute aid equitably at the local level; this resulted in people having to rely on informal networks and local actors to manage the relief distribution process.

The Nepali government's Post-Disaster Needs Assessment (PDNA), undertaken by the National Planning Commission (NPC, 2015) within two months of the earthquake estimated that 700,000 additional Nepalis would be pushed into poverty between 2015 and 2016 as a result of the earthquake (Lam and Kuipers, 2019; Regmi, 2016). The PDNA indicated the scale of damage by the earthquake in preparation for the 'Toward a Resilient Nepal' international donor conference on 25 June 2015. The outcome of the conference was positive, with pledges amounting to about US$4.4 billion aid from the international community. However, the international community placed stress on the importance of transparency during the relief effort, and reiterated the need for 'political stability' – a reference to expedite Nepal's agreement on the delayed constitution – as preconditions for the release of funds.

At the time of the earthquake Nepal was in the process of drafting a constitution, but the process had been stalled due to fundamental disagreements over a number of key contentious issues. Nepal was already burdened by poverty and poor infrastructure (Chughtai, 2015; Regmi, 2016) and in most of the 14 districts affected by the earthquake, an estimated 41 per cent of the damaged houses belonged to lower caste ethnic *Dalits* and indigenous groups, 26 per cent to families headed by women and 23 per cent to elderly people (Amnesty International, 2017; Nougaret and Danuwar, 2016). The poor had therefore been heavily affected and the painfully slow reconstruction process did not help matters (Mishra and Adhikari, 2019; von Meding et al., 2017). The prolonged period of uncertain political transition, reflected in the failure to draft a constitution, a vacuum in local governance, and a recurrent pattern of political brinksmanship and power-brokering kept Nepal perpetually on the edge of a political cliff. The Nepali government's preoccupation with its own political issues made it significantly unprepared to deal with any kind of disaster. Furthermore, the politics of drafting the constitutional led to inaction and partisanship that delayed much of the reconstruction, despite billions of dollars of aid provided by the international community (Mishra and Adhikari, 2019). Delays to the large-scale infrastructure projects created cascading effects, slowing building reconstruction work. The Nepali economy suffered tremendous losses due to reconstruction delays. Eventually, the 'fast-tracked' constitutional process led to a new statute passed on 20 September 2015.

In addition, the disbursement of funds had been made dependent on recipients achieving progress on building code compliance (von Meding et al., 2017). The government adopted an 'owner-driven' reconstruction model, which required proof of land ownership as a condition to qualify for the reconstruction grant (Amnesty International, 2017). However, it has been estimated that up to 25 per cent of Nepal's population is landless or near-landless (Amnesty International, 2017). As a result of this owner-driven reconstruction approach, tens of thousands of people have been denied reconstruction aid (Amnesty International, 2017) (see Figure 5.5). This reconstruction model has widened inequalities and worsened the vulnerabilities of the marginalised, including the landless, women, *Dalits* and other caste-based and ethnic minorities (Amnesty

Figure 5.5 People living in poor households in informal settlements, such as the one pictured, were unable to access reconstruction support, forcing them to live in damaged housing in vulnerable conditions

International, 2017; Lam and Kuipers, 2019). The whole recovery and reconstruction process has been characterised by a low level of community inclusion and participation (Lam and Kuipers, 2019).

At the beginning of the reconstruction process, the government of Nepal promised plans that would achieve 'established well-planned, resilient resettlements and a prosperous society' and to 'build back better' (Lam and Kuipers, 2019, p. 322). However, it appears that the hope of rebuilding Nepal into a resilient society is gradually being undermined by the process (Lam and Kuipers, 2019). Instead of building back better, Nepal's earthquake reconstruction process is being driven by private property ownership (Amnesty International, 2017). Yet property rights have been undermined by inefficiency in the judicial system, witnessed by substantial corruption and political influence (Heritage Foundation, 2019; World Economic Forum, 2015).

National Reconstruction Authority

Setting up a National Reconstruction Authority (NRA) to oversee the reconstruction progress was officially constituted in August 2015 under a special ordinance.

The NRA announced that US$7.1 billion was needed for the recovery and reconstruction (Mishra and Adhikari, 2019; Sharma et al., 2018). However, the NRA was automatically dissolved due to the failure of the government to have it approved by the Legislature within the required time frame. Infighting among the political parties over the control of the NRA was one of the key reasons for not meeting the time frame set for ratifying the ordinance to establish the NRA. A new bill on the NRA was tabled in the parliament in September 2015 and was passed after three months in January 2016 with a new CEO. This chain of events stalled the reconstruction efforts for a significant period of time.

Senior government officials challenged the political appointment of the new NRA Chief Executive Officer (CEO), who was perceived to be a junior officer in the bureaucratic hierarchical system. Technocrats in the government openly protested the appointment and refused joining the NRA, resulting in deficiencies in technical and governance capacities. This led the NRA to function while it was in the process of building itself, that is, drafting policies and directives while also implementing them. This resulted in an absence of a strong, integrated directive during the initial stages of reconstruction, leading to decisions being made at the district level that were open to interpretation. The changes to policies at the district level – specifically adjusting definitions of 'affected persons' and conditions of eligibility to aid – led to discrepancies in the recording of the number of affected households impacting on most affected marginalised groups, particularly single women and disabled people (Lord and Moktan, 2017).

Despite the US$4.4 billion in aid for the reconstruction from the international community (von Meding et al., 2017; World Economic Forum, 2015), only 16 per cent of the amount has been disbursed to affected households three years after the earthquake (Mishra and Adhikari, 2019). While the owners of 60 per cent of the 90,000 damaged private houses have received the first instalment of reconstruction grants, this only represents 17 per cent of the construction aid (Mishra and Adhikari, 2019; von Meding et al., 2017). Official statistics at the end of June 2018 revealed that the rebuilding rate had reached 21.7 per cent, representing 169,526 households out of the 781,733 affected households entitled to government aid (Lam and Kuipers, 2019).

Key institutions for reconstruction in Nepal

The Ministry of Home Affairs acts as the National Focal Agency on Disaster Management. The Ministry is responsible for rescue and relief work and was the lead agency responsible for implementation of any of the acts governing disaster resilience in Nepal (Dangal, 2011; Regmi, 2016). Prior to 2017, disaster-related initiatives were governed under the Natural Calamity (Relief) Act, 1982 (Dangal, 2011) and the Local Self Governance Act 1999 (Shrestha and Pathranarakul, 2018). However, in 2017 a newly approved Disaster Risk Reduction and Management Act (2017) was enacted (Nepal et al., 2018). Under this act, disaster resilience responsibility is shared among central, regional, district and local committees (Dangal, 2011; Nepal et al., 2018). Other acts that regulate aspects of disaster resilience include the Soil and Watershed Conservation Act (1982) and the Building Act (1998) (Nepal et al., 2018).

Under the Building Act (1998), the Department of Urban Development and Building Construction (DUDBC) assumed the responsibility for developing and updating the National Building Code (Nepal et al., 2018). However, the implementation of the National Building Code and other requirements of the Act is the purview of local governments (Nepal et al., 2018). The 2015 Earthquake recovery and reconstruction efforts in Nepal are being spearheaded by the National Reconstruction Authority (NRA) (Mishra and Adhikari, 2019; Shrestha and Pathranarakul, 2018). Non-governmental organisations (NGOs) have also been instrumental in the recovery and reconstruction (Jones et al., 2014; Regmi, 2016). A Post Disaster Recovery Framework (PDRF) (2016–2020) was formulated by the NRA as a guiding framework to manage the reconstruction process that adopts an 'owner-driven' approach to disaster recovery (Pokharel et al., 2018). However, there have been concerns that the NRA is understaffed and not working effectively, and that the five-year time frame for reconstruction indicated by NRA could further worsen the plight of the affected (Nepali Times, 2016; Pokharel et al., 2018). Instead of enabling reconstruction, the authority became occupied with 'procedural issues and seems to be more interested in controlling the purse strings than in fast-tracking reconstruction' (Nepali Times, 2016, 11–17 March 2016 #799 Editorial). The Post Disaster Recovery Framework itself was questioned, with critics saying that it 'completely ignored the procedural inequities that might be created because of the pre-existing socio-economic disparities' (Pokharel et al., 2018, p. v).

Policy aspects of disaster resilience: pre- to post-2015 earthquake

Over the years, the government of Nepal has developed various legal and institutional arrangements to plan for, reduce and manage disaster risks. The government has attempted to develop a well-structured functional institutional set-up from the centre to local levels that can have positive outcomes in managing disasters. Following are some of the important legal and policy frameworks (c.f. Ministry of Home Affairs (2019) National Policy for Disaster Risk Reduction):

a) Natural Calamity (Relief) Act 1982;
b) Local Self Governance Act 1998; Building Act 1998;
c) National Building Code 2004;
d) National Strategy for Disaster Risk Management 2009;
e) Climate Change Policy 2011, Land Use Policy 2012;
f) Water Induced Disaster Management National Policy 2015;
g) National Reconstruction and Rehabilitation Policy 2015;
h) National Disaster Response Framework 2013;
i) Basic Guideline related to Settlement Development, Urban Planning and Building Construction 2016;
j) National Urban Development Strategy 2016;

k) Local Government Operation Act 2017;
l) Disaster Risk Reduction and Management Act 2017;
m) National Policy for Disaster Risk Reduction, 2018.

The last two acts, Disaster Risk Reduction and Management Act (2017) and National Policy for Disaster Risk Reduction (2018), are major legal arrangements that were recently approved and enacted by the Constitution of Nepal. Nepal et al. (2018) suggest that the Disaster Risk Reduction and Management Act (2017) is a milestone document for managing disasters in Nepal. This act, as indicated earlier, overrides the Natural Calamity (Relief) Act (1982); also as a piece of policy, for the first time, it considers the different stages of the disaster risk management cycle – preparedness, response and rehabilitation, and mitigation. More importantly, the Government of Nepal has recognised the need for a strong institutional set-up to address disaster risk reduction through the coordination of different levels of government. The 'Natural Disaster Relief Committees', established at different levels of administration as per the law, and various agencies and institutes of the Government of Nepal are all important institutional arrangements supporting resilient communities. 'The National Council, Executive Committee, National Disaster Risk Reduction and Management Authority and Disaster Management Committees at Province, District and Local levels are new institutional arrangements as per the Disaster Risk Reduction and Management Act 2017' (The Government of Nepal, 2018).

Building construction practices and regulations in Nepal

The earlier post-1988 earthquake damage assessment revealed that destruction of buildings was mainly due to poor construction, a low level of awareness of earthquake-resilient construction technology and non-existence of clear building code requirements. This led to the National Building Code (NBC) Development Project, formulated through the Department of Buildings assisted by UNDP (UNDP Project NEP/88/054). The original 20-volume draft code with four levels of application was prepared in 1994 and presented regulations and guidelines for buildings to control and improve the seismic performance of buildings to meet an acceptable level of safety. The NBC incorporates four different aspects of building types according to the complexity of design and construction:

1. *International state-of-the-art*: Applicable to large building structures. These include buildings with complex designs usually found in developed countries. The structures must comply with existing international state-of-the-art building codes. However, they must also meet Nepalese requirements with respect to minimum design load and configuration.
2. *Professionally engineered structures*: Buildings designed and constructed under the supervision of engineers, buildings with a plinth area more than 93 square metres, buildings having more than three storeys, buildings with a span more than 4.5 metres and buildings with irregular shapes fall into

this category. It covers public buildings like hospitals, meeting halls, factories, multi-storey buildings and larger residential buildings.

3. *Mandatory rules of thumb*: Buildings of restricted size designed by simple rules of thumb. The mandatory rules of thumb can be used in the case of buildings that meet certain criteria related to height, number of storeys and floor area. They be used by non-experienced builders and junior engineers to apply for a building permit and applied in the construction process. Buildings of a plinth area less than 93 square metres, less than three storeys and having a span less than 4.5 metres, and regular-shaped buildings designed and constructed by technicians in areas where professional engineers are not available. These rules were revised after the 2015 Earthquake.

   The documents that cover the aspects of mandatory rules of thumb are:

   - Mandatory Rules of Thumb Reinforced Buildings with Masonry Infill (NBC 201)
   - Mandatory Rules of Thumb Reinforced Buildings without Masonry Infill (NBC 205)
   - Mandatory Rules of Thumb Load Bearing Masonry (NBBC 202).

4. *Focused Guidelines*: For remote and rural buildings where control is impractical, the guidelines focus on changes required to make buildings seismic-resistant, which are not subjected to modern quantitative analysis and rational design considerations. They include earthen buildings and low-strength buildings (unfired masonry, mud mortar, rubble, and dry stones) and Guidelines for Remote Rural Buildings (low strength masonry/earthen buildings) constructed by local masons not more than two storeys high. Documents included in this section are:

   - Guidelines for Earthen Resistant Building Construction: Earthen Buildings (NBC 204)
   - Guidelines for Earthen Resistant Building Construction: Low strength Masonry (NBC 203).

The enforcement of the initial NBC was enacted through the 'Building Act 1998'. However, due to technical difficulties the enforcement was not very successful. The Department of Urban Development and Building Construction (DUDBC) added further three volumes to the code. The entire set consisting of 23 volumes was approved by the Government of Nepal in 2003. The Code contains provisions pertaining to architectural, structural, mechanical and electrical requirements and the objective is to protect the public by accounting for the influence of local conditions on construction.

The Building Act 1998 was amended in 2007, followed by the Building Regulation 2008, to improve implementation of the codes. Although the codes were applicable to the whole country and expected to serve as a legally binding document across all municipalities, it was not mandatory for village development committees that existed before recent local-level restructuring in Nepal. Only a few municipalities adopted the building code before the 2015 Gorkha Earthquake. Lalitpur sub-metropolitan city was the first municipality in Nepal to implement the NBC in the Building Permit Process through a declaration on the Earthquake Safety Day held on 16 January 2003. After the 2015 earthquake, four volumes of the codes were revised to reflect the new experiences; in addition, the National Building Codes (NBC) became mandatory for every municipality and is now enforceable. The main municipalities and also the village municipalities are bound to follow the NBC for new construction which includes the extensive reconstruction after the Gorkha Earthquake.

Problems of using Mandatory Rules of Thumb (MRT) for constructing buildings

Until the development of the Nepal National Building Code in 1994, building construction in Nepal did not adhere to earthquake-safe building construction practices (Karki, 2019; Subedi et al., 2016). However, the impetus for earthquake-safe building came after the earthquake of 1988 (Subedi et al., 2016). This earthquake triggered the incentive for stakeholders to work towards safer buildings in Nepal. These efforts culminated in the publication of the Nepal National Building Code (NBC) in 1994 (Subedi et al., 2016).

The NBC was published with the aim of improving the structural quality of future buildings in order to minimise the loss of life during earthquakes as a result of structural failure (Ministry of Physical Planning and Works, 1993). The NBC specified the Mandatory Rules of Thumb (MRT) to provide ready-to-use guidelines for owner-builders in Nepal constructing buildings with a reinforced concrete (RC) frame, including ordinary residential buildings of up to three storeys (Ministry of Physical Planning and Works, 1993). Generally, three types of buildings are under the remit of MRT: (1) RC buildings with masonry infill; (2) RC buildings without masonry infill; and (3) load-bearing masonry buildings (Subedi et al., 2016). The intention of the MRT was to attain minimum seismic safety requirements. It was designed to be used by mid-level technicians, such as overseers and draughtspersons, with limited training in structural design of buildings (Ministry of Physical Planning and Works, 1993).

Effective implementation of building codes can prevent the collapse of buildings and save the lives of thousands of people during earthquakes (Guragain et al., 2018; Hall et al., 2017). However, the implementation of the NBC in Nepal has been a challenge (Ahmed et al., 2018). Karki (2019) and Subedi et al. (2016) concurred that new buildings are still being erected without compliance with the earthquake safety guidelines enshrined in the NBC. The MRTs in particular have not been broadly patronised and conventional approaches to construction, which are mostly unsafe, are still in use (Subedi et al., 2016).

One barrier to compliance with building codes is that constructions adhering to MRT have been found to be ten times more expensive than those using conventional unsafe approaches (Subedi et al., 2016). Not surprisingly, more than 95 per cent of buildings that collapsed during the 2015 earthquake were built from low-strength masonry allowed by the MRT (World Economic Forum, 2015). The MRT implementation has also faced challenges due to the complex building-permit process and delays in getting the permits (Karki, 2019). It is generally reported that building codes are not being effectively implemented due to lack of building inspections, corruption, absence of local government mechanisms, as well as an overburdened judicial system (Ahmed et al., 2018; World Economic Forum, 2015).

'Safe building codes' implementation after the 2015 earthquake

Earthquakes pose a serious risk in Nepal and disaster-resilient building codes are essential to reduce impacts, an issue that came into sharp focus after the massive earthquakes in 2015. Although recovery and reconstruction from the earthquakes has been slow, the government and other stakeholders have been taking the implementation of building codes seriously.

Awareness about the need for safe building codes

The Earthquake Safety Day, observed on January 16th since 1999, commemorates and reflects on the 1934 earthquakes; this event has spread some level of awareness about earthquake safety. After the 2015 earthquake experience, most Nepali people did not want to compromise the safety of their buildings. Such events are important because people tend to forget about the need to prepare for future events during the time between earthquakes.

Three key groups of people are evident when it comes to adoption and implementation of safe building codes. The first group can be described as 'carefree'. An earthquake expert mentioned this group of 'people are not careless, but they are not careful also – they are very carefree.' This group of people have been constructing buildings without adequate awareness of the building laws and building quality standards. They also hold inaccurate conceptions about safe buildings: for example, they believe that just using steel or some form of solid material in the construction will make their homes earthquake-resistant. With more information and education, this group may potentially consider appropriate adoption of safe building codes. Implementation of building codes in areas where traditional settlements are common is difficult, as it is widely perceived that traditional construction technology has stood the test of time. The NBC has focused on these communities and provided more detail on the benefits of new or modern technology, such as reinforced concrete, but any improvements to traditional construction are not fully covered by the codes. Furthermore, the highly dense traditional settlements have certain construction details and characters which the NBC has not been able to incorporate in detail.

The second group can be characterised as having a belief in safe building codes, but they tend to place their faith in the market or the external agencies and institutions to build safe buildings for them. This group is vulnerable on several accounts: (a) Negotiations for lowering construction prices without being conscious of the implications, leading to contractors relaxing the quality of construction; (b) Contractors consciously ignoring safe building practices to reduce cost as there are limited compliance checks from government bodies; and (c) Relying on contractors or trade persons who are not fully informed or qualified to apply the building codes – often these are trades persons and engineers who have worked in the field for a long time and do not keep track of the changes to codes or regulations, thus they are unaware of what they do not know. Therefore, training is very important to keep such trades persons and professionals educated on constructing safe buildings.

The third group is aware and knowledgeable of safe construction, but they do not follow or practice it due to different constraints. One such major constraint is limited urban space. Even if this group is not willing to compromise safety, the tension between the small land sizes in urban areas and the demand for more building space compels them to compromise on safe building practices. As an example, following setback rules of the building, by-laws established by the local authority will make the already small houses on small-sized urban land very small and as a result people do not want to follow safe building practices. In rural areas, this group may simply not have the resources, such as money or skills, to enable them to follow safe building practices. As this group is aware and knowledgeable of the safe building practices and making a conscious decision not to follow the practices so as to maximise their opportunities, it is difficult to make them adopt safe building practices. The difficulty with the process of obtaining building permits or approval also impacts on the willingness to engage with building codes.

Meanwhile, awareness of safe building codes led some homeowners of buildings not damaged by the earthquakes to voluntarily dismantle the higher levels of their house – those which were self-built and non-compliant to building codes. As an example, some owners of three- to five-storey buildings voluntarily dismantled one or two storeys to make the building safer. Despite such practices, the level of safe building awareness is not high or consistent across the nation.

Training and quality improvement initiatives

The interview results indicated that awareness and skills to implement the NBC were mostly lacking in Nepal prior to the 2015 Gorkha Earthquake. In general, technicians and engineers lacked sufficient knowledge for designing urban high-rise buildings. It was also observed that new engineering graduates had limited understanding of the NBC, even though it is an important aspect of building disaster resilience.

After the 2015 earthquake, training relating to the NBC was improved to educate engineers and trades persons at different levels. The government has also developed a number of guidelines, construction specifications, pamphlets and prototype designs. Training on safe building construction is provided to different levels of technicians such as engineers, sub-engineers, assistant sub-engineers and masons. Under the governmental programs more than a hundred thousand masons and contractors received training ranging in duration from one week to two months. The training of masons was seen as one of the successful strategies to promote voluntary compliance by many municipalities. Some agencies also used megaphones and media channels for dissemination of information regarding the NBC. The Dharan and Thimi municipalities conducted homeowner counselling every week for those who were interested in applying for a building permit.

There were also a number of training programmes for owner-builders. Such builders comprise a significant part of the housing reconstruction, which makes it difficult to monitor the quality of reconstruction. As it is not possible to stop owner-built buildings, it is important that owner-builders are trained. Some women trained in masonry after the earthquake formed cooperatives to construct their homes themselves (see Figure 5.6). In these cooperatives, a group of 20 to

Figure 5.6 Owner-built housing is common in Nepal and masonry training for women allows better gender balance in this construction process

25 people get together to take turns building the house of each cooperative member. Although this was a significant empowerment initiative, there were questions about the adequacy of a two-week masonry training course to design and build earthquake-resilient buildings. The two-week course for women was effective when they were working under the directions of long-time practising masons where the house owner only helped as a labourer, not as a proper mason.

The enforcement of building codes in all municipalities is ensured through the building permit system. The building permit is provided for the construction of a building only when the design and drawings of the building fulfil the NBC provisions and by-laws; a building completion certificate is thereafter provided to demonstrate that the building has been constructed as per the design and drawings. Municipalities such as Dharan have also implemented a reward-and-punishment system where contractors who do not build according to NBC-compliant design and drawings will lose their licence to work in their municipality. Furthermore, upon demonstration of good work, the municipality rewards contractors during different events such as the Earthquake Safety Day. The same municipality has also introduced supervision of construction work with a small additional charge. They also provide incentives to owners (50 per cent discount in fee) when they apply for a permit to retrofit their existing building.

Despite the wide training initiatives, some of the onsite reconstruction done by owner-builders and masons in the rural areas – especially in residential buildings – lacks NBC compliance, as was evident in the detail of lapping length of reinforcement bars, stirrups and ties. The ambiguity arising from the initial version of the Mandatory Rules of Thumb (MRT) also created problems. Some of the issues with the MRT were revised after the 2015 earthquake, but it has proven difficult to communicate such changes to rural construction workers or owner-builders.

Changing labour market conditions and professional support

Only a few large contractors operate in Nepal and most of them demonstrate professionalism. However, a large part of the construction sector is run by small businesses or sole masons. The construction skills training provided by the government has had a positive impact on local construction. However, most people (anecdotal evidence suggests 40 to 50 per cent) who were trained left Nepal after some period of time to work in other countries, such as in the Middle Eastern Gulf region. In Nepal, exploitation of the workers and the temporary nature of work, unpredictable employment patterns and underpayment by contractors has not helped the financial security of trained masons. Therefore, they sought other opportunities overseas with more stability. Although the pay levels have improved over the years, the labour drain is ongoing. This situation demands continuous training to keep up with the skill drain. However, as the overseas income remittance is a key part of the economy (Karnik, 2015), this skill drain may also provide a positive economic outcome through income remittance and development of workforce in Nepal.

Although the training programme for engineers served the urban areas very well, most of the rural areas did not have adequately trained engineers to support quality assurance and code-compliant building practices. A government engineer's salary package was less attractive than that of the similar level of engineers working in international NGOs. The unsatisfactory income demotivated some government engineers, leading to very limited engagement with the public, specifically in rural areas. Moreover, the large volume of reconstruction work commissioned by NGOs changed the prices and wages structures in the market because of the higher payment for products and services to achieve progress with the reconstruction work; this left the people who do not receive any financial support with difficulty in rebuilding or moving forward with recovery. On the other hand, increases in wages created higher levels of demand for masons and construction workers, thus creating local employment opportunities.

The construction labour market was mobile in attracting skills from non-affected districts to affected areas and also across the border from India. Because of the urgency of the reconstruction and associated demand, the locals welcomed the external labour supply. As the government has placed the entitlement for reconstruction subsidies based on a tight timeline and staged payments, to meet the timeline for subsidy entitlements, beneficiaries had to rely on any form of available labour supply. The mobility of labour also helps with dealing with the skill drain, as mentioned above.

Institutional behaviours: facilitating or controlling reconstruction

As part of ensuring reconstruction compliant to safe building codes, the NRA prepared about 20 types of standard off-the-shelf residential designs. Reconstruction subsidies were only offered to people who used one of the NRA-approved designs. If anyone wanted to change the standard design, it had to be undertaken by qualified engineers and approved by the local authority to receive the subsidies. There was urgency among some groups of beneficiaries to get the reconstruction completed as quickly as possible to receive government subsidies that were linked to the construction progress milestones. The NRA's aim was to expedite the reconstruction process while ensuring that it was earthquake-safe. However, evidence suggests that many of the beneficiaries have only received the first or second instalment of the subsidies and the rest of the instalments did not progress as fast as they should have. Although the experts believe that, technically, it is not a good idea to rush the reconstruction, the practices adopted by the NRA in terms of linking subsidy instalments with milestones was generally viewed as a reasonable initiative.

There was some criticism of the NRA being control-minded with a 'big stick' rather than a facilitator. The NRA was also criticised for being too top-down in its approach – in early March 2016, with a brief notice the NRA halted the reconstruction of destroyed or damaged private homes undertaken without its approval, affecting 2.5 million survivors. However, some argue that if the NRA did not control and be prescriptive with the design, people would construct non-compliant structures themselves with the subsidies, which is contrary to the objective of enabling resilience.

The NRA and other related government agencies were criticised for the lack of or ambiguous communication to the public. Any institutional approach for post-disaster reconstruction needs to be designed and implemented with proper communication. The gaps in communication made people unsure of what they were supposed to do to proceed effectively with the reconstruction of their properties. This was exacerbated by a lack of technical support, specifically inadequate engineers, and some of the available engineers did not have an adequate understanding of the reconstruction process to be able to provide the required reconstruction support.

Financing reconstruction: key strategic and operational challenges

Insurance for buildings, particularly residential buildings, is not commonly practiced in Nepal. High premiums, specifically for a large bulk of old buildings, was a hindrance for insurance uptake. Moreover, a lack of trust in insurance companies made people avoid taking out insurance. People tend to be unsure whether insurance companies will pay the claims after an adverse event, or whether it will be a long and difficult process to get the claims processed. Social groups such as those comprised of marginalised and low-income people are the most vulnerable as they do not have the understanding or means for adopting an insurance scheme.

Additionally, as land tenure is a condition for receiving the government reconstruction subsidy, this disallows a substantial segment of the population who have unclear landownership status. To be eligible for insurance and to receive any government subsidy, disaster survivors need to prove their ownership of land. Many survivors had difficulty in supplying land title documents, especially people who had built informally (Nougaret and Danuwar, 2016). Moreover, the land had to fit certain parameters set in the reconstruction guidelines to be eligible for subsidies – for example, the requirement that a plot suitable for reconstruction should have 6 metres of road clearance at some places or 4 metres in rural areas. Affected people who did not fit into the stringent guidelines were left out of the government post-disaster support system. As many people could not satisfy the stringent conditions set in the national reconstruction guidelines, the guidelines were subsequently revised by the Cabinet to ease the grants approval process. The revisions also allowed – after due processing by the NRA – provision of resettlement land to beneficiaries with earthquake-damaged houses that had been built informally on encroached public land, forest land or other government land and had no other means to access land.

Resettlement of communities, with the aim of avoiding habitation in high earthquake- and landslide-risk areas or allocating alternative land to address legal title issues, is a complex task and brings problems of its own. Such endowments require significant planning and resources and can take many years to complete. However, sometimes communities themselves were left to organise or arrange relocations. Most people did not want to move from their current habitats to resettle in a new place because of their existing networks and employment. Also, for the new settlers, to be accepted by the existing communities was a challenge. The relocated communities face language, social class, economic and psychological challenges.

Institutional arrangements, resourcing and urban transformation

The National Disaster Report (2017) identified a number of key priority issues that Nepal needs to address for a disaster resilient future, and all them require improvements to the institutional structure, directly or indirectly. The Government of Nepal has addressed some the key institutional set-up concerns by establishing Natural Disaster Relief Committees at different levels of administration, and various agencies as suggested in the Disaster Risk Reduction and Management Act 2017, to assist capacity building at all levels of the government for disaster risk reduction, preparedness, response and recovery. The following are some of the areas where resourcing will make a significant impact to building resilience:

- *Empowering and resourcing institutions*: Setting up appropriate institutions is a key step in the right direction for building disaster resilience. However, for institutions to deliver positive outcomes they need to be adequately resourced and empowered. The provincial and local governments, as key institutions in building disaster resilience, need to develop effective leadership roles in disaster risk reduction and management. As part of a constitutional change to expand its institutional network, Nepal moved from 58 municipalities in 2014 to 278 municipalities in 2017, and these municipalities need to be strengthened and supported to develop the capacity to foster safer building construction. They also need to work in alignment with broader national institutional directions. However, ongoing resourcing of the key institutions to deliver their prime objectives remains a challenge.
- *DRR-CCA discourse*: Ensuring allocation of adequate recognition and funding for the governance of Disaster Risk Reduction (DRR) and Climate Change Adaptation (CCA) integration at all levels of institutions is essential for achieving disaster resilience. Institutions need to be actively engaged in the DRR-CCA discourse to make a long-term impact. Instituting a practice of risk-informed development and mainstreaming DRR and CCA into sectoral development planning should be important policy considerations. DRR-CCA is also a platform for facilitating gender-responsive disaster resilience. It also forms part of the wider sustainable development discourse and informing land use planning and building codes.
- *Land use planning and building codes*: It is viewed by experts in Nepal that the municipality should consider land use planning and safe building at the core of its role, and establish a land use planning-cum-building permit system. However, the building development process and land use planning are not very well integrated in Nepal, leading to significant governance gaps in the development process. This needs support from all levels of government. Municipalities in some places have only one engineer working with infrastructure development and housing. With

commitments to reviewing drawings, and working on construction sites involving frequent travel and site visits, it is very difficult to effectively monitor the quality of construction. Based on the population and construction workload, both human and financial resources need to be increased in these municipalities.

- *Information systems and communication*: Disaster resilience is strongly linked to building a diverse and robust knowledge base and communicating the knowledge to make people understand their risks and take appropriate measures to reduce the risks. Setting up an effective disaster information management system as a knowledge and information hub can improve the access to information for at-risk communities.

Conclusion

Nepal is exposed to a number of natural hazards including earthquakes, floods and landslides. As a developing country, Nepal experiences severe impacts of these natural hazards on its economic development. Nepal was underprepared to deal with the scale of devastation wrought by the 2015 Gorkha Earthquake disaster. Political instability, shadows of the past insurgency, weak institutions and corruption, along with the constraints of its mountainous geography impeded the earthquake recovery and reconstruction process. The impact of the earthquake was overwhelming to the nation as a whole. The immediate aftermath of the earthquake was a difficult moment for Nepal; engaging with and meeting the conditions of international donors with an unstable political environment was challenging. However, despite setbacks the Nepal Parliament passed a new constitution and established the National Reconstruction Authority (NRA) to oversee the reconstruction process.

In Nepal, the critical role and importance of building codes in developing disaster resilient communities is acknowledged at the institutional level. Nepal has continually developed a set of National Building Codes to suit different circumstances since the early 1990s. Developing and implementing building codes in a developing country with a large rural population, difficult geological conditions and income disparity has many challenges that are economic, political and cultural. The Nepal building codes and regulations were set under several categories with the view of covering all demographics and building types; the Mandatory Rules of Thumb indicates a flexible approach, recognising the limitations of low-income and rural communities in adhering to stringent codes.

After the 2015 earthquake, a number of new institutions were developed to strengthen the governance of disaster resilience initiatives. The newly approved 'Disaster Risk Reduction and Management Act 2017' consolidated and replaced many old acts. The Ministry of Home Affairs assumed the role of the National Focal Agency on Disaster Management and the lead agency responsible for implementation of the acts relating to disaster resilience in Nepal. Although the implementation of the National Building Code and the requirements of any other relevant act is in the purview of local governments (Mishra and Adhikari, 2019; Shrestha and Pathranarakul, 2018), the NRA has reshaped the way safe buildings are constructed by using approved designs for reconstruction. However, some of the initiatives of the NRA drew criticisms that instead of enabling reconstruction, for example, the authority was occupied with procedural issues and used controlling behaviours.

The key issues impacting the adoption of building codes include, firstly, the lack of available professional engineering and trades skill in the market. Some of the fresh graduates lacked understanding of the NBC. The lack of awareness among some of the professionals of the importance of NBC and its role in disaster resilience contributes to the perpetuation of Nepal's vulnerability to earthquakes. Trained masons leaving the country creates constant skill shortages in Nepal. This can be addressed by continuous training and improving, and also by addressing the gender balance in masonry skills. Secondly, inadequate institutional capacity posed a barrier to effective building code implementation. Even though since the 2015 earthquake new institutions have been created, they lack adequate institutional capacity in terms of skilled technical human resources and also financial resources. The institutional context suffers from a range of issues – lack of an effective reward-and-punishment system; political pressure to issue building permits or completion certificates, although the design and construction of a building lacks compliance to building codes; gaps in the building permit system as the monitoring of the foundation is missing during the process of construction; consistency and lack of clarity in building codes and building quality assurance; and importantly, a lack of strong political will to enforce building codes.

Furthermore, large stocks of old buildings exist, which were not built according to building codes – retrofitting them is challenging and expensive (see Figure 5.7). At the community level, ensuring good quality of materials is often difficult. Most of the interview respondents pointed out that awareness of all the stakeholders is the basic and key aspect to help encourage voluntary compliance at the community level. In Nepal, where most houses are owner-built, it would be fruitful to develop incentive mechanisms for the house owners to encourage NBC compliance. This needs to be accompanied by a proper system in the municipality with a less bureaucratic process for building approvals. In essence, Nepal needs to find ways to formalise and standardise the construction of traditional technology in informal settlements to increase the adoption of building codes. This requires more professional and research engagement, perhaps drawing from the experience of the wider South Asian region, to conduct evidence-based policy and code development.

Thus, some key strategic approaches to overcoming the challenges and improving the compliance of NBC include: capacity development at all levels; increasing the awareness of different stakeholders; more evidence-based practices involving research institutions and academia; the need to find solutions on retrofitting the large bulk of existing buildings, many of old stock; and better incentive schemes designed to engage different stakeholders.

Figure 5.7 There is a large stock of old and vulnerable buildings in Nepal, such as these in Kathmandu, being propped up as a short-term measure to prevent collapse

References

Ahmed, I., Gajendran, T., Brewer, G., Maund, K., von Meding, J., Kabir, H., Faruk, M., Shrestha, H.D. and Sitoula, N.R. (2018) *Understanding the Opportunities and Challenges of Compliance to Safe Building Codes for Disaster Resilience in South Asia* (technical report). Kobe: Asia Pacific Network for Global Change Research (APN).

Amnesty International. (2017) *'Building Inequality': The Failure of the Nepali Government to Protect the Marginalised in Post-Earthquake Reconstruction Efforts*. London: Amnesty International.

Bhattacharjee, A., Anadón, J., Lohman, D., Doleck, T., Lakhankar, T., Shrestha, B., Thapa, P., Devkota, D., Tiwari, S., Jha, A., Siwakoti, M., Devkota, N., Jha, P. and Krakauer, N. (2017) 'The Impact of Climate Change on Biodiversity in Nepal: Current Knowledge, Lacunae, and Opportunities'. *Climate*, 5(4), pp. 80–96.

Bohara, A.K., Mitchell, N.J. and Nepal, M. (2006) 'Opportunity, Democracy, and the Exchange of Political Violence: A Subnational Analysis of Conflict in Nepal'. *The Journal of Conflict Resolution*, 50(1), pp. 108–128.

Chamlagain, D. and Gautam, D. (2015) 'Seismic Hazard in the Himalayan Intermontane Basins: An Example from Kathmandu Valley, Nepal' in Shaw, R. and Nibanupudi, H.K. (eds.) *Mountain Hazards and Disaster Risk Reduction*. Tokyo: Springer, pp. 73–103.

Chughtai, S. (2015) *Nepal Was Made Vulnerable by More Than Its Violent Geology*. The Guardian. Available at: www.theguardian.com/commentisfree/2015/apr/26/nepal-earthquake-poverty-infrastructure-aid (accessed 16 September 2019).

Daly, P., Ninglekhu, S., Hollenbach, P., Barenstein, J.D. and Nguyen, D. (2017) 'Situating Local Stakeholders within National Disaster Governance Structures: Rebuilding Urban Neighbourhoods Following the 2015 Nepal Earthquake'. *Environment and Urbanization*, 29(2), pp. 403–424.

Dangal, R. (2011) *Country Profile: Nepal*. Kathmandu: ADRC.

Deraniyagala, S. (2005) 'The Political Economy of Civil Conflict in Nepal'. *Oxford Development Studies*, 33(1), pp. 47–62.

Dixit, A. (2013) *Climate Change in Nepal: Impacts and Adaptive Strategies*. World Resources Report. Available at: www.wri.org/our-work/project/world-resources-report/climate-change-nepal-impacts-and-adaptive-strategies (accessed 16 September 2019).

Dixit, A.M., Shrestha, S.N., Guragain, R., Pandey, B.H., Oli, K.S., Adhikari, S.R., Acharya, S.P., Jimee, G.K., Upadhyaya, B.K., Sangachhe, S.B., Shrestha, N., Pradhan, S., Dhungel, R., Khatiwada, P., Baskota, A., Poudel, A., Tandingan, M., Mathema, N.U., Parajuli, B., Basyal, G.K., Chaudhary, S., Bhatta, G.R. and Marasini, N. (2018) 'Risk Management, Response, Relief, Recovery, Reconstruction, and Future Disaster Risk Reduction' in Gautam, D. and Rodrigues, H. (eds.) *Impacts and Insights of the Gorkha Earthquake*, vol. 1. San Diego: Elsevier Science, pp. 95–134.

Dixit, A.M., Yatabe, R., Dahal, R.K. and Bhandary, N.P. (2013) 'Initiatives for Earthquake Disaster Risk Management in the Kathmandu Valley'. *Natural Hazards*, 69(1), pp. 631–654.

Economist. (2017) *Nepal: Political Structure*. The Economist. Available at: http://country.eiu.com/article.aspx?articleid=1385795722&Country=Nepal&topic=Summary&subtopic=Political+structure (accessed 27 September 2019).

Gaire, S., Castro Delgado, R. and Arcos González, P. (2015) 'Disaster Risk Profile and Existing Legal Framework of Nepal: Floods and Landslides'. *Risk Management and Healthcare Policy*, 8, pp. 139–149.

Gautam, D., Rodrigues, H., Bhetwal, K.K., Neupane, P. and Sanada, Y. (2016) 'Common Structural and Construction Deficiencies of Nepalese Buildings'. *Innovative Infrastructure Solutions*, 1(1), p. 1.

Global Facility for Disaster Risk Reduction (GFDRR). (2019) *Nepal*. World Bank. Available at: www.gfdrr.org/en/nepal (accessed 3 October 2019).

Guragain, R., Pradhan, S., Maharjan, D.K. and Shrestha, S.N. (2018) '13 – Building Code Implementation in Nepal: An Experience on Institutionalizing Disaster Risk Reduction in Local Governance System' in Shaw, R., Shiwaku, K. and Izumi, T. (eds.) *Science and Technology in Disaster Risk Reduction in Asia*. London, United Kingdom: Academic Press, pp. 207–220.

Hagelsteen, M. and Burke, J. (2016) 'Practical Aspects of Capacity Development in the Context of Disaster Risk Reduction'. *International Journal of Disaster Risk Reduction*, 16, pp. 43–52.

Hall, M.L., Lee, A.C.K., Cartwright, C., Marahatta, S., Karki, J. and Simkhada, P. (2017) 'The 2015 Nepal Earthquake Disaster: Lessons Learned One Year On'. *Public Health*, 145, pp. 39–44.

He, L. (2019) 'Identifying Local Needs for Post-Disaster Recovery in Nepal'. *World Development*, 118, pp. 52–62.

Heritage Foundation. (2019) *2019 Economic Freedom Index, Nepal*. Available at: www.heritage.org/index/country/nepal (accessed 15 September 2019).

Index for Risk Management (INFORM). (n.d.) *Country Profile: Nepal*. European Commission Disaster Risk Management Knowledge Centre. Available at: https://drmkc.jrc.ec.europa.eu/inform-index/Countries/Country-Profile-Map (accessed 3 October 2019).

International Federation of Red Cross and Red Crescent Societies (IFRC). (2015) *Emergency Plan of Action (EPOA) Nepal: Earthquake*. Kathmandu: IFRC.

International Federation of Red Cross and Red Crescent Societies (IFRC). (2019) *Nepal: Monsoon Floods and Landslides*. Kathmandu: IFRC.

International Monetary fund (IMF). (2018) IMF Staff Completes 2018 Article IV Mission to Nepal: Press Release No. 18/473. Available at: www.imf.org/en/News/Articles/2018/12/12/pr18473-nepal-imf-staff-completes-2018-article-iv-mission (accessed 7 August 2019).

Jones, S., Oven, K.J., Manyena, B. and Aryal, K. (2014) 'Governance Struggles and Policy Processes in Disaster Risk Reduction: A Case Study from Nepal'. *Geoforum*, 57, pp. 78–90.

Jones, S., Oven, K.J. and Wisner, B. (2016) 'A Comparison of the Governance Landscape of Earthquake Risk Reduction in Nepal and the Indian State of Bihar'. *International Journal of Disaster Risk Reduction*, 15, pp. 29–42.

Karki, T.K. (2019) 'What Did the 2015 Earthquake Tell Us about What the State of Earthquake Resilience in Kathmandu Metropolitan City Was?'. *International Journal of Disaster Resilience in the Built Environment*, 10(2/3), pp. 188–202.

Karnik, M. (2015) *Billions of Dollars from Overseas Workers Will Be Key to Rebuilding Nepal's Ravaged Economy*. Quartz India. Available at: https://qz.com/india/392011/billions-of-dollars-from-overseas-workers-will-be-key-to-rebuilding-nepals-ravaged-economy/(accessed 07 September 2019).

Lam, L.M. and Kuipers, R. (2019) 'Resilience and Disaster Governance: Some Insights from the 2015 Nepal Earthquake'. *International Journal of Disaster Risk Reduction*, 33, pp. 321–331.

Lord, A. and Moktan, S. (2017) *Uncertain Aftermath Political Impacts of the 2015 Earthquakes in Nepal*. Conciliation Resources. Available at: www.c-r.org/accord/nepal/uncertain-aftermath-political-impacts-2015-earthquakes-nepal (accessed 7 September 2019).

Matambo, S. and Shrestha, A. (2013) *Nepal: Responding Proactively to Glacial Hazards*. World Resources Report Case Study. Washington, DC: World Resources Report.

Ministry of Home Affairs. (2018) *Nepal Disaster Report, 2017: The Road to Sendai*. Kathmandu: Government of Nepal.

Ministry of Home Affairs. (2019) *National Policy for Disaster Risk Reduction Policy 2018*. Kathmandu: Government of Nepal.

Ministry of Home Affairs (MoHA). (2015) *Nepal Disaster Report 2015*. Kathmandu: Ministry of Home Affairs (MoHA) and Disaster Preparedness Network-Nepal (DPNet-Nepal). Available at: www.drrportal.gov.np/uploads/document/329.pdf (accessed 2 October 2019).

Ministry of Physical Planning and Works. (1993) *Nepal National Building Code (NBC 205, 1994): Mandatory Rules of Thumb Reinforced Concrete Buildings without Masonry Infill*. Kathmandu: Government of Nepal.

Ministry of Population and Environment. (2016) *Nationally Determined Contributions of Nepal*. Kathmandu: Ministry of Population and Environment.

Mishra, S.R. and Adhikari, B. (2019) 'Planetary Health in Nepal's Post-Earthquake Rebuilding Agenda: Progress and Future Directions'. *The Lancet Planetary Health*, 3(2), pp. 55–56.

National Planning Commission (NPC). (2015) *Nepal Earthquake 2015: Post Disaster Needs Assessment*. Kathmandu: Government of Nepal.

Naujoks, J. (2016) *Building Back Better or Restoring Inequalities? Gender and Conflict Sensitivity in the Response to Nepal's 2015 Earthquakes*. International Alert. Available at: www.international-alert.org/sites/default/files/Gender_ConflictSensitivityNepalEarthquakes_EN_2016.pdf (accessed 2 October 2019).

Nepal, P., Khanal, N.R. and Prasad, B. (2018) 'Policies and Institutions for Disaster Risk Management in Nepal: A Review'. *The Geographical Journal of Nepal*, 11, pp. 1–24.

Nepali Times. (2016) 'Deconstruction Authority: The Government Should Stop Trying to Be a Control Freak, and Open up the Space for Earthquake Survivors to Easily Rebuild Their Homes'. *Nepali Times*, 17 March 2016.

Neupane, S.P. (2015) 'Immediate Lessons from the Nepal Earthquake'. *The Lancet*, 385(9982), pp. 2041–2042.

Nougaret, A. and Danuwar, R.P. (2016) *Housing, Land and Property Issues in Nepal and Their Consequences for the Post-Earthquake Reconstruction Process*. Kathmandu: CARE Nepal.

Oxfam and Humanitarian Accountability Monitoring Initiative. (2019) *Fighting Inequality in Nepal: The Road to Prosperity*. Oxford: Oxfam International.

Pokharel, T. and Goldsworthy H.M. (2017) Lessons Learned from the Nepal Earthquake 2015. *Australian Journal of Structural Engineering*, 18(1), pp. 11–23, DOI: 10.1080/13287982.2017.1309818

Pokharel, T., Manandhar, M.D., Dahal, A., Chalise, B., Bhandari, R. and Kharel, T.P. (2018) *Political Economy Analysis of Post-Earthquake Reconstruction in Nepal: An Assessment of Emerging Role of Local Governments*. Kathmandu: Nepal Administrative Staff College and The Asia Foundation.

Quora. (2018) *What Is the Present Federal Structure of Nepal and What Are the Significances*. Quora. Available at: www.quora.com/What-is-the-present-federal-structure-of-Nepal-and-what-are-the-significances (accessed 2 October 2019).

Rafferty, J.P. (2015) *Nepal Earthquake of 2015*. Encyclopaedia Britannica. Available at: www.britannica.com/topic/Nepal-earthquake-of-2015 (accessed 2 October 2019).

Rana, B.S. (2015) *Nepal Ko Maha Bhukampa (1990)* (in Nepalese). Kathmandu: Brahma Shumsher Rana.

Regmi, K.D. (2016) 'The Political Economy of 2015 Nepal Earthquake: Some Critical Reflections'. *Asian Geographer*, 33(2), pp. 77–96.

Reid, K. (2018) *2015 Nepal Earthquake: Facts, FAQs, and How to Help*. World Vision. Available at: www.worldvision.org/disaster-relief-news-stories/2015-nepal-earthquake-facts (accessed 3 October 2019).

Rose, L.E., Matinuzzaman, Z.P., Karan, P. and Proud, R.R. (2019) *Kingdom of Nepal, Nepāl Adhirājya*. Encyclopaedia Britannica. Available at: www.britannica.com/place/Nepal/Climate (accessed 27 September 2019).

Ruszczyk, H.A. (2018) 'A Continuum of Perceived Urban Risk – From the Gorkha Earthquake to Economic Insecurity'. *Environment and Urbanization*, 30(1), pp. 317–332.

Sharma, K. (2006) 'The Political Economy of Civil War in Nepal'. *World Development*, 34(7), pp. 1237–1253.

Sharma, K., Kc, A., Subedi, M. and Pokharel, B. (2018) 'Challenges for Reconstruction after Mw7.8 Gorkha Earthquake: A Study on a Devastated Area of Nepal'. *Geomatics, Natural Hazards and Risk*, 9(1), pp. 760–790.

Shrestha, B. and Pathranarakul, P. (2018) 'Nepal Government's Emergency Response to the 2015 Earthquake: A Case Study'. *Social Sciences*, 7(8), p. 127.

Shrestha, S. (2015) *Deconstructing Reconstruction: Rebuilding Destroyed Homes Has Been Unnecessarily Delayed by Lack of Money, Masons and Clear Government Policy.* Nepali Times. Kathmandu. https://archive.nepalitimes.com/article/nation/earthquake-survivors-waiting-to-rebuild,2566.

Shrestha, S., Gautam, D.R., Gorkhaly, G.P., Adhikari, B.R., Shrestha, Y., Gautam, M.S., Sitaula, N. and Mahat, B. (2018) *Nepal Urban Resilience Project (NURP) Scoping Study.* Kathmandu: Adventist Development and Relief Agency.

Subedi, J., Ghimire, R.M., Neupane, R.P. and Amatya, S. (2016) 'Cost Difference of Buildings in Kathmandu Constructed with and without Earthquake Safer Features'. *International Journal of Disaster Resilience in the Built Environment*, 7(5), pp. 444–459.

Thapa, D. and Ramsbotham, A. (2017) *Two Steps Forward, One Step Back the Nepal Peace Process.* Conciliation Resources. Available at: www.c-r.org/downloads/Accord-26-Nepal-WEB_0.pdf (accessed 2 October 2019).

The Government of Nepal. (2018) *National Policy for Disaster Risk Reduction*. Kathmandu: Ministry of Home Affairs.

von Meding, J., Shrestha, H.D., Kabir, H. and Ahmed, I. (2017) *Nepal Earthquake Reconstruction Won't Succeed Until the Vulnerability of Survivors Is Addressed.* The Conversation. Available at: www.theconversation.com/nepal-earthquake-reconstruction-wont-succeed-until-the-vulnerability-of-survivors-is-addressed-87335 (accessed 2 October 2019).

World Bank. (2019) *World Bank Country and Lending Groups: Country Classification.* Available at: https://datahelpdesk.worldbank.org/knowledgebase/articles/906519-world-bank-country-and-lending-groups (accessed 2 October 2019).

World Economic Forum. (2015) *Building Resilience in Nepal through Public-Private Partnerships. Global Agenda Council on Risk & Resilience*. Available at: www3.weforum.org/docs/GAC15_Building_Resilience_in_Nepal_report_1510.pdf (accessed 27 September 2019).

World Population Review. (2019) *World Population by Country.* Available at: http://worldpopulationreview.com/ (accessed 26 September 2019).

The 2017 flooding and landslides
Post-disaster policy implementation in Sri Lanka

6 Framing accountability

Policy implementation following floods and landslides in Sri Lanka

Introduction

Within this chapter, disaster management policy is explored in terms of implementation following the devastating flooding and landslide disasters that affected Sri Lanka during 2017. A country located within the Indian Ocean, it remains an island separated from the Indian peninsula by the Palk Strait. In this respect, Sri Lanka shares no land borders and lies isolated from the mainland; thus, remains vulnerable to hazard events, particularly those of a hydro-meteorological origin (Foundation for Environment, Climate and Technology, 2017).

Sri Lanka frequently experiences disaster events such as tsunamis, storms, floods and landslides; concurrently, many regions are exposed to severe drought conditions (Ministry of Disaster Management, 2019a; Ministry of National Policy and Economic Affairs and Ministry of Disaster Management, 2017). More recently, the intensity and frequency of climate-induced disasters has increased across the country; whilst rapid urbanisation, combined with unauthorised development activities, has played a further role towards the high level impacts experienced by local communities from disaster related events (FECT, 2017; MNPEA and MDM, 2017).

Over the past decades, floods and landslides have become more prevalent, exacerbated by climate change (Mawilmada et al., 2010). In general, these disaster events are caused by the south-west and north-east monsoon seasons, resulting in heavy rains that lead to flooding and subsequently landslides, given the vulnerability of the inundated land (Disaster Management Centre, 2017; FECT, 2017). The devastating flooding and landslide disasters that afflicted Sri Lanka during 2017 demonstrated the destructive nature of such events, where communities continue to remain vulnerable and resilience remains negligible.

Although Sri Lanka has formal legislation governing disaster management and a plethora of agencies responsible for execution of such policy, there remains minimal post-disaster evaluation of implementation – as a result, such policy may often lack operational effectiveness. Therefore, this chapter aims to explore the area of implementation to offer guidance for future policy direction. Given the unremitting disasters across the region, subsequent property devastation and the significant loss of life, policy evaluation can present an opportunity to understand current practice that can direct future change, ultimately assisting communities to build resilience.

The content of this chapter is presented in two parts. First, the chapter provides an overview of the country and identifies the primary hazard conditions that generate such significant vulnerability. The regulatory system that has subsequently evolved to counter disaster impacts is then explored, as this system encapsulates the policy environment and subsequent implementation actions. Second, the chapter considers the 2017 flooding and landslide disasters with a focus on Ratnapura – a major city in the Sabaragamuwa Province – considered one of the most affected regions, in order to provide a deep understanding of the strengths and challenges of policy implementation. The discussion incorporates key stakeholder experiences to explore mechanisms for change, with the aim of contributing to future policy decisions and supporting vulnerable communities in building resilience.

The research involved two key methodological approaches. First, a comprehensive review of literature was conducted, including both academic and grey sources, such as media commentaries. Second, fieldwork for data collection was undertaken. The main data collection methods involved stakeholder interviews and focus groups, in conjunction with observations and photographic evidence. All data was analysed through content analysis, leading to grouping of themes, coding and cross-referencing to interpret and present the findings of this chapter.

Disaster overview: 2017 flooding and landslide disaster event

The 2017 Sri Lanka floods and subsequent landslide disasters resulted in significant devastation across Sri Lanka. Over 200 people were killed and around 70 people remain missing. There was significant damage to property, vital major infrastructure and health services, as well as destruction of livelihoods – around 600,000 people were displaced as a result of the floods (MNPEA and MDM, 2017).

The severe weather events experienced across the country are often a result of the South Asian monsoon season: the South-west Monsoon and the North-east Monsoon (Department of Meteorology, 2019). In the case of the 2017 flood and landslide event, the Disaster Management Centre (2017, p. 1) explained:

> Southwest monsoon was activated over Sir Lanka from 24 May 2017 and a very low upper air wind convergence was formulated over the eastern sea of Sri Lanka which was absorbing westerly winds. As a result of this, heavy rains were received on 25 May to the South-western watersheds in the country. Large amount of rains were received within 12 hours in SW regions including Namunuthanna (619mm) of, Bulathsinghela (419mm), Morawaka (406mm) and Walasmulla (437mm) leading riverine-floods of the Kalu, Nilwala and Gin rivers.

The torrential rains and winds associated with the monsoon resulted in severe flooding and ultimately destabilised the ground, resulting in catastrophic landslides. The country was inundated by heavy rains for more than 12 hours that eventually led to the flooding of the Kalu, Nilwala and Gin rivers and their tributaries (MNPEA and MDM, 2017). The impact on the river basins resulted in urban flooding across coastal areas, and in some cases substantial water diversions that caused flooding across many other localities. In some regions, water levels above 5 metres were experienced for many days. With such high water levels and destabilisation of lands, a total of 35 major landslides occurred, mainly focused around the south-west regions. It is worth noting that many districts experienced additional landslide events; however, these were considered more minor in nature (MNPEA and MDM, 2017).

In many cases, landslides were attributed to a combination of vulnerability of saturated lands and slope instability – notably, a result of urban land practices. The District of Ratnapura was subjected to a high number of these landslides (MNPEA and MDM, 2017), with the source generally attributed to a combination of unauthorised development and recent infrastructure activities. The MNPEA and MDM (2017) reported that many recent road developments had inadequate drainage which provided a foundation for water collation, creating land destabilisation and water diversion channels.

With the severe weather conditions, the landslide events exacerbated the flooding devastation, resulting in a significant number of deaths: across the country there was an estimated 219 deaths. In total it has been estimated that around 900,000 people were impacted by the 2017 disaster events (DMC, 2017; MNPEA and MDM, 2017). Around 15 districts were severely affected, with a total of 504 relief camps or safety centres established to assist more than 100,000 people (MNPEA and MDM, 2017). The DMC Rapid Impact Assessment Report, prepared in June 2017, illustrates the degree of impact on the population due to the torrential rains experienced during the May event (see Table 6.1). Although the centres were generally focused upon those immediately affected by the flood and landslide events, they also became a haven for those facing potential landslides in regions identified as unstable or high risk.

The 2017 event also had significant impacts upon business activities, vital health and public infrastructure services. From an economic perspective, the disasters had adverse implications across many regions with agriculture, trade and industry affected. Major agricultural areas used for tea, rubber, coconut and paddy plantations were destroyed through saturation and subsequent siltation impacts. In total it was estimated that more than 300,000 people had their livelihoods negatively impacted (DMC, 2017; MNPEA and MDM, 2017).

Table 6.1 People affected by 2017 floods and landslides (Source: DMC, 2017, p. 1)

Description	*Amount (No. of people)*
Severely affected	70,000
Moderately affected	465,000
Women directly affected	290,000
Children directly affected	138,000
Houses impacted	20,792

Once the initial monsoonal conditions abated, the destruction of property through landslides continued and many communities remained in a position of vulnerability. Many people had to rebuild with no foreseeable or immediate future income to support such activities, and on the same land that remains at the mercy of the next hazard event. Given the country is subjected to frequent hazards, the government has introduced a plethora of policy and initiatives to assist with community preparedness and decrease vulnerability. However, it is important to review and evaluate these systems and processes from a policy perspective, particularly following such events. Therefore, the 2017 event – that was both recent and highly destructive – provides the case study for policy implementation evaluation in this chapter from which the learnings, strengths and challenges can help direct future positive change.

Country overview

The island of Sri Lanka is located in the Indian Ocean (see Figure 6.1) and has a range of topographical features, with the morphology of the terrestrial environment including low-lying coastal areas, a mountain range in the southern central part of the island and associated valleys, ridges and plateaus (Department of Meteorology, 2019; MNPEA and MDM, 2017). Otherwise, most regions across the country are considered quite flat. Given the diversity in relation to topography, combined with the isolated location within an ocean, this impacts upon wind patterns, rainfall, temperature and associated climatic conditions (DoM, 2019). In terms of land mass, Sri Lanka has an overall land size of around 65,000 square kilometres (Zubair et al., 2006), is approximately 432 kilometres long at its widest point and maintains a maximum width of approximately 224 kilometres (Peiris and Arasaratnam, 2019).

Urbanisation, employment and transportation

With regard to population, Sri Lanka had approximately 20.7 million people in 2017 (MNPEA and MDM, 2017); in 2019 this has increased to well above 21 million (World Bank, 2019; World Population Review, 2019). The cities and towns continue to rapidly expand near the rivers bordering the main cities; however, a significant proportion of the population continues to reside in rural areas often due to the association with employment activities. Economically, Sri Lanka is considered a low to middle income country with inequity evident across the regions. In this respect, the often-labelled 'fragile' economy has resulted in one quarter of the population being identified as 'nearly poor' (MNPEA and MDM, 2017) and subsequently remaining exposed to sudden economic variation or, as termed by the United Nations, 'shocks' (United Nations Development Programme, 2016). Unfortunately, one of the primary shocks that produces significant disruption to the economy across multiple levels relates to hazards and the resulting disaster events (UNDP, 2016).

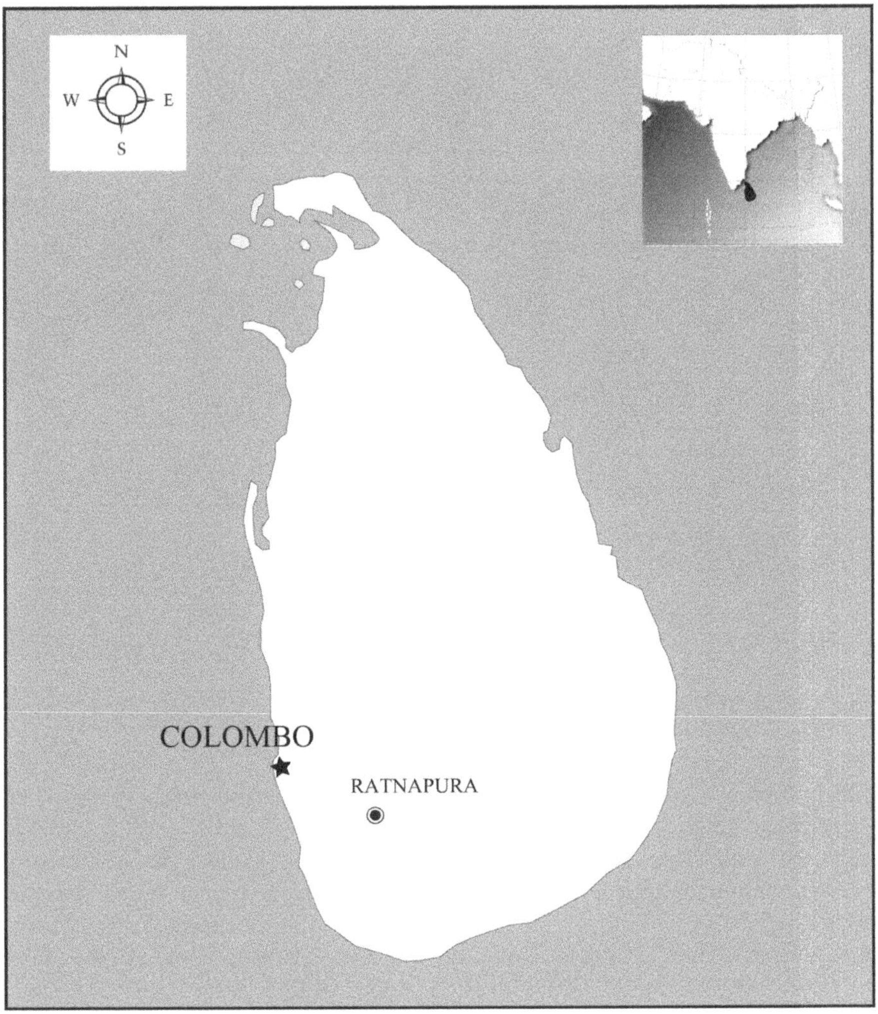

Figure 6.1 Geographical location of Sri Lanka

Urbanisation, particularly around the coastal regions, brings the total housing stock in Sri Lanka to an estimated 20.36 million (MNPEA and MDM, 2017); a quarter of this population reside in the capital, Colombo, and its wider metropolitan area (World Population Review, 2019). Given the current rate of growth, the Ministry of Housing has identified that approximately 30,000 new houses are constructed across the country each year (MNPEA and MDM, 2017) rendering the construction market a large contributor to Gross Domestic Product

Figure 6.2 A view from the rapidly urbanising city of Galle

(GDP). Figure 6.2 provides an example of the urban environment for one of the major cities, Galle.

In line with cities, rural areas have continued to expand due to the vast array of employment opportunities presented. In this respect, rural activities are a significant contributor to the economy with the agriculture, forestry and fisheries sectors accounting for around 7.9 per cent of the GDP (MNPEA and MDM, 2017). Rice – a staple of the local diet – has been identified as an important crop nationally, involving around 1.8 million farming families (MNPEA and MDM, 2017). However, the most significant agricultural sector is the fishing industry, accounting for around 1.3 per cent of the GDP (Central Bank of Sri Lanka, 2017). Overall, the country produces around 80 per cent of the food requirements domestically to support the population (MNPEA and MDM, 2017; United Nations World Food Programme, 2017).

Sri Lanka maintains a road network that consists of expressways managed by the Road Development Authority, as well as national, provincial and local roads. In addition, there are a number of local 'unclassified' roads that technically remain outside the scope of those categories formerly identified, often due to the inadequate nature of their construction. In general, privately owned vehicles are popular with a range of additional public transportation options including buses and hire vehicles. The country is also serviced by rail for both the movement of passengers and goods (MNPEA and MDM, 2017).

Environment

Sri Lanka is a country that exhibits a range of different environments, from mountains in the southern central area to low-lying coastal areas, including wetlands. From a climate perspective, there are four distinct seasons: first inter-monsoon (March to April), south-west monsoon season (May to September), second inter-monsoon season (October to November) and north-east monsoon season (December to February) (DoM, 2019). Interestingly, in a country plagued by consistent hydro-meteorological conditions, drought is common and becoming more prevalent (FECT, 2017; Zubair et al., 2006).

From an environmental perspective, urbanisation – particularly construction activities (see Figure 6.3) – as well as commercial plantations and mining have significantly affected the ability of the land to manage and mitigate flooding and landslide events. Anthropogenic impacts, in particular dumping of waste, have significantly altered waterways and therefore impacted on their ability to also assist with management of disaster events through the creation of obstructions and minimising natural filtration. Typically, wetlands play an important

Figure 6.3 Settlements such as this one in Ratnapura town that have been built on riverbanks are exposed to the impacts of flooding

role in water detention during flood events and as such changes to these environs through construction activities have the potential to increase impacts of flooding and landslide events. Additionally, these impacts extend into other water bodies such as rivers, that are often subjected to pollution through waste dumping on adjacent lands (MNPEA and MDM, 2017).

Disaster risk profile

As noted in the former section on the country profile, Sri Lanka is a country subjected to frequent hazards that often bring forth devastation and destruction in terms of loss of life, property and economic vulnerability. Flood events are common, and in the ten-year period from 2005, floods impacted more than 60 per cent of the population. In 2016 alone, it is estimated that around half a million people were affected by flooding and landslides; the devastation resulted in the deaths of over 90 people with more than 100 reported as missing (Ministry of National Policies and Economic Affairs and Ministry of Disaster Management, 2016). Landslides, often a direct result of flooding, are responsible for the most human casualties: in the 2017 event, 176 deaths were a direct result of landslide events (MNPEA and MDM, 2017). The effects of disasters are felt by many communities and unfortunately have a flow-on effect impacting housing, livelihoods, transportation, as well as vital health services, education and water and sanitation sectors.

The impacts upon local communities following the effects of the flooding and landslides events in 2017 were apparent with five regions severely affected: Galle, Hambanthota, Kalutara, Matara and Ratnapura (MNPEA and MDM, 2017). Low density rural areas experienced the most impact as they have little means of access to others forms of livelihood when agricultural land is damaged. Subsequently, farming communities are often the most vulnerable and experience severe hardship during these events; this is exacerbated often by a loss of housing. Following the 2017 event, it was evident that many poorly constructed houses in rural areas received significant damage.

Given the frequent occurrence of hazards and subsequent disaster events, the Disaster Management Centre and Ministry of Disaster Management (2019) developed a hazard profile of Sri Lanka: nine of the most predominant hazards were identified to assist with future planning and risk minimisation. The following discussion considers each of these hazards in the context of the complexity of the environment, and serves to assist in understanding the risk of exposure.

Coastal erosion

Coastal erosion has increasingly impacted communities across Sri Lanka; this is of high significance, given the high degree of the population that reside within these waterfront areas. Erosion is economically significant on a national level: more than 40 per cent of GDP relates to coastal activities (Wickramaarachchi, 2013). Essentially, the island is influenced by the hydrodynamics of the Indian

Ocean as well as supply of sediment from terrestrial areas – in particular river alluvial. Continual erosion and accretion across the coast is part of the functioning coastal system; however, a number of actions can occur to alter the balance between these. Coastal erosion is exacerbated during the monsoon season (South-west and North-east), in particular by the monsoonal waves (Wickramaarachchi, 2013). Overall, the island is affected by ongoing hazards such as those associated with monsoons and the more immediate erosion impacts from tsunamis and storms.

Anthropogenic development has both increased erosion and exposed land uses and infrastructure to further risk. Human activities including construction activities along coastal areas as well as extraction of sand cause additional impact and risk. Furthermore, these development related activities have created significant erosion regions near communities such as in the western, south-western and southern coastal areas (Wickramaarachchi, 2013).

Drought

Drought, due to a significant and continual reduction in rainfall, is generally considered a 'slow-onset' (De Silva and Punyawardena, 2013, p. 48) hydro-meteorological hazard. Across Sri Lanka the climate, in particular rainfall levels, is affected due to: the country being an island surrounded by ocean, it is located in the warm tropics, the central hills are subject to monsoon wind streams and the land masses to the north and north-west (De Silva and Punyawardena, 2013). Climate change has been identified as a factor that has significantly increased drought risk (Mawilmada et al., 2010), to the point that it is now considered one of the most frequently occurring disasters. The changes in rainfall, or lack thereof, are due to a number of climatic conditions ranging from a high pressure air stream that traverses over the dry landmass of neighbouring India before reaching Sri Lanka, through to decreased rainfall coming from the Bay of Bengal causing disruption to weather systems (De Silva and Punyawardena, 2013).

Regardless of the cause, the influence of droughts in Sri Lanka has been experienced for many years, affecting a broad geographic area and therefore many communities. In a way, drought is insidious, given its often subtle appearance compared to other hazards such as flooding where impacts are often immediately apparent. De Silva and Punyawardena (2013) explains that without the immediate 'notable' devastation evident with so many other disasters, it is often difficult to quantify the impacts from this form of event. Nonetheless, agricultural and plantation operations – important economic sources – may be significantly affected by drought.

Floods

Within Sri Lanka, floods are acknowledged as the most frequent type of hazard. In fact, many areas of the country may be considered vulnerable to flooding as

Figure 6.4 A flooded area in Ratnapura district

they receive high rainfall, are of low elevation and are located close to waterways (Senaratne, 2013). Ratnapura and Kalutara are often identified as the most vulnerable localities, and the picture in Figure 6.4 shows the degree of flooding that may be experienced. Frequently the country is accosted by severe flooding events and the national hazard profile has categorised these as riverine, flash and localised events (Senaratne, 2013). However, it is important to note that flooding may also be a result of flood control works when, for example, these implemented measures malfunction or collapse: possibly as a result of former flooding events.

Over time, a range of satellite images in conjunction with land-based surveys have been employed to prepare inundation maps – to help understand the hazard profile. Additionally, they serve as the mechanism by which measures associated with flood mitigation may be implemented. In many cases these relate to structures to control the flood waters and incorporate flood levees and flood ways. Although there is much evidence to support the effectiveness of structural mitigation measures, unfortunately, these methods bring an economic burden particularly in relation to ongoing maintenance activities. Therefore, non-structural measures serve as an important mechanism to reduce vulnerability: options that focus on improved planning operations associated with education, in conjunction with appropriate land use management (Senaratne, 2013). Typically, this form of mitigation may consider land zoning to ensure appropriate use, through to awareness campaigns associated with preparedness and warning.

Landslides

While landslides – soil and rock movement downslope due to gravity (United Nations Educational, Scientific and Cultural Organization, 2017) – occur across the landscape of Sri Lanka, their impact is increasing due to people relocating and settling into more heavily sloping areas (Weerasinghe and Jayatissa, 2013). Furthermore, anthropogenic land use activities including agriculture and settlements within vulnerable land areas can be seen to exacerbate the risk of landslide as they contribute to changing the natural environment.

Landslide occurrence is increased primarily during seasonal heavy rainfall during the monsoon and inter-monsoon seasons. Weerasinghe and Jayatissa (2013) identified that a significant 20 per cent of the country is considered to be at risk of landslides, affecting ten districts. Figure 6.5 provides an example of the type of topography across the high risk areas and highlights land instability issues facing these regions. Hazard mapping has assisted with reducing risk from a land use planning perspective, as well as with the relocation of at-risk communities (Weerasinghe and Jayatissa, 2013). However, landslide events continue to occur, impacting many regions and resulting in a heavy toll due to loss of lives. Damage to buildings and property renders people homeless and incapacitates vital services and infrastructure. Across more rural areas communities generally suffer hardship and economic impacts through destruction of lands associated with agricultural and plantation practices. Overall, as landslides have

Figure 6.5 A landslide in Ratnapura district

increased, these disasters present a continual risk nationally resulting in death, loss of property and livelihood.

Lightning

Lightning strikes are commonly between clouds, known as a cloud flash, or between the cloud and the ground, known as a ground flash (Kumarasinghe, 2013). Lightning strikes may be affected by a host of climatic, geographical and environmental conditions such as 'longitude, latitude, altitude, soil conditions, topography and vegetation' (Kumarasinghe, 2013, p. 106). Across Sri Lanka, the strikes are quite profound and include primary and secondary impacts, both with devastating consequences. Primary impacts are associated with injury and death and the significant disruption to industry and infrastructure services. Secondary impacts across the region refer to those 'charges' that spark fires and explosions. A staggering 51 people perished from lighting strikes in 2011; the following year, 43 people were killed (Kumarasinghe, 2013), demonstrating the devastating death toll from this often unrecognised hazard.

In many cases the impact of a strike will also affect industry, such as electricity and communication systems, given the potential for fire and property damage (Kumarasinghe, 2013); this subsequently may restrict operational activities and livelihood in flow-on effects. Although lightning strikes may cause disruption to many services, they may have dire consequences to vital health services such as hospitals and during emergency events with regard to warning, communication and rescue services. Ultimately, the lighting strikes that occur in Sri Lanka are a significant concern and warrant inclusion as a hazard given that they may cause injury, death and property destruction.

Sea level rise

The sea level is predicted to rise by around 59cm by 2100 (Intergovernmental Panel on Climate Change, 2007; Ratnayake, 2013) due to global warming. There are two primary causes of sea level rise: generally the first is associated with thermal expansion due to heat and the second from increased mass of the ocean resulting from glacial melts (Ratnayake, 2013). The issue for Sri Lanka is that it is susceptible to climate change impacts (Ministry of Environment, Climate Change Secretariat, 2010) as it is an island with low-lying coastal areas. Therefore, given the expanse of shoreline, it is predicted that these areas will be negatively affected by climate change with rising water levels and coastal erosion (Mawilmada et al., 2010; Ratnayake, 2013), rendering communities vulnerable and in many cases displacing people from their homes.

However, the impacts from a rising sea level go beyond those residing within low-lying coastal areas, as communities may be forced inland and to higher ground. An increase in density across these regions, and potentially within 'high risk' locations, subjects communities to landslides and other hazard events that will bring further challenges nationally. Urbanisation brings another threat and enhances the impacts of climate change through vegetation removal, changing land use and activities such as those associated with sand mining. At present, hazard modelling provides a means by which to predict future shoreline changes, and enables government's ability to prepare and plan for what may be considered an unpredictable future.

Storm surge

A traditional storm surge is due to 'high winds pushing on the ocean's surface' (Wijethunga, 2013a, p. 143), generally associated with cyclone events that affect water levels and impact coastal areas that may result in flooding of low-lying areas. Cyclones that form in the Bay of Bengal are the most significant for Sri Lanka, while those that form in the Arabian Sea often have less influence although may still produce storm surges. Interestingly, many cyclones form within the Bay of Bengal and the Arabian Sea, yet few of them make it far enough to have a significant effect on the people in Sri Lanka. It is estimated that between 1891 and 2000 around 450 cyclones formed over the Bay of Bengal, yet only 4 per cent made landfall in Sri Lanka (Wijethunga, 2013a).

During a cyclonic event the country is affected by the combination of high winds and heavy rain as well as dangerous storm surges. Similar to sea level rise modelling, storm surge hazard mapping has been undertaken to identify coastal areas at risk of inundation to assist with future planning, primarily associated with early warning systems (Wijethunga, 2013a). However, cyclones and subsequent coastal inundation have resulted in a significant injury and death toll over the years, as well as damage to housing and infrastructure services, presenting a devastating situation particularly to coastal communities.

Tropical cyclones

Cyclones, formed over warm water, may gather energy while traversing over the ocean with a temperature greater than approximately 26°C. They occur in tropical environs and are associated with strong winds, heavy rain and storm surges that have a significant impact on local communities and ecosystems within low-lying coastal areas (Wijemannage, 2013). Sri Lanka, due to its coastal location, is considered to be on the 'periphery of the tropical cyclone belt' (Wijemannage, 2013, p. 177); as such, it experiences cyclonic events that generally have less impact. Regardless, history has shown that cyclones have occurred that resulted in significant loss of life and damage to property and infrastructure across the region.

As highlighted within the discussion on storm surges, many tropical cyclones form over the North Indian Ocean – Bay of Bengal and Arabian Sea – yet only a small percentage make landfall (Wijethunga, 2013a). However, the devastation and destruction from those tropical cyclones that reach land with full force may be significant. Most cyclones that make landfall tend to affect the east coast of the country (Wijemannage, 2013); the loss of life and property destruction that may occur during these events is generally attributed to heavy rainfall that results in flooding in addition to strong winds that contribute to the devastation.

Tsunami

A tsunami wave usually presents with a long wavelength – often hundreds of kilometres long – with a shallow amplitude formed under sea by an earthquake or volcanic eruption (Wijethunga, 2013b). When the tsunami nears landfall the wave height or amplitude increases due to the low profile of the shoreline (Wijethunga, 2013b), and thus causes significant physical damage to coastal areas. Unfortunately, a tsunami generally does not occur in isolation: it is usually followed by a series of waves, compounding the devastation caused by the initial wave.

Although Sri Lanka is subjected to few tsunamis, those experienced have often had truly disastrous consequences. The most notable tsunami to affect the country was on 26 December 2004. The result of an earthquake within the Sumatra–Andaman subduction zone (Wijethunga, 2013), the 2004 Indian Ocean quake and tsunami caused significant devastation. Wijethunga (2013b) reported that the event had a death toll of over 35,000 with more than 20,000 injured and around 100,000 dwellings destroyed. The damage to property, infrastructure and subsequently livelihoods was immense, displacing many communities. Although considered an infrequent event in Sri Lanka, the severe impacts from tsunamis necessitate their inclusion in the risk profile for the country.

Impacts of urbanisation

Many disasters, including flood and landslide events, are exacerbated by urbanisation and land use activities. Typically, the ability of the natural environment to cope with disasters has been affected over the years, as many of the urban and agricultural practices have reshaped the natural environment: resulting in destabilisation of local lands and modification of waterways, rendering them vulnerable to flooding and subsequent landslides. Common examples include clearing in riparian zones with subsequent reduction in riverbank stability, development activities that have impacted natural detention and filtration across wetlands, and of course increased siltation in waterways. The continual desire to cultivate land for agricultural purposes creates additional land change and thus vulnerability.

Regions most affected by floods and landslides often present the highest levels of development with concentration of urban population and associated buildings, infrastructure and employment generating land use activities within low-lying areas. Many development activities are not controlled, may be considered illegal, adding to the impacts of authorised developments. Development, contradictory development controls and limited governance or enforcement all provide impact environmental changes that contribute to increased impacts and vulnerability (MNPEA and MDM, 2017).

Given the frequency of hazards that affect Sri Lanka, the MDM presented nine categories of hazard for Sri Lanka: disaster risk profiles, as summarised above. Through this mechanism, risk resilience activities can be focused across the categories in an attempt to build capacity and mitigate impact. However, identification of risks and implementation of mitigation planning is often insufficient to warrant change. In many cases, formal regulatory policy is needed to evoke change and direct future activities. Sri Lanka maintains a hierarchal regulatory system to guide disaster resilience activities; this will now be explored to provide the backdrop to the case study for this project.

Governance: disaster resilience policy and process

From a National perspective, Sri Lanka presents a complex administrative structure to guide disaster resilience-related operations during times of emergency. The Sri Lanka Disaster Management Act No. 13 of 2005 (Sri Lanka) provides a framework to govern activities related to disaster resilience (Ashraf et al., 2017; Lalitha et al., 2015). The Act is a regulatory policy – legislation – enacted by the government and is considered:

> an Act to provide for the establishment of the national council for disaster management; the disaster management centre; the appointment of technical advisory committees; the preparation of disaster management plans; the declaration of a state of disaster; the award of compensation and formatters connected therewith or incidental there to.
>
> (DMA No. 13 2005 (Sri Lanka); MDM, 2019b)

The National Council for Disaster Management (NCDM) was established to provide overarching directives for management of disasters at a national level (MDM, 2019c): the governing agency responsible for disaster resilience activities. The Ministry of Disaster Management came into effect as the governmental agency to lead activities for disaster-related directives. Nationally, both agencies are responsible for administering the provisions of the Disaster Management Act (DMA). The roles and responsibilities of the MDM are significant and incorporate disaster mitigation, response and recovery as related to formulation of policies and programmes. In addition, the MDM are accountable for the formulation of two high level plans: the National Disaster Management Plan and the National Emergency Operation Plan. Their powers extend to projects associated with early warning systems, rescue operations, as well as research (MNPEA and MDM, 2017).

The MDM has four reporting institutions: the Disaster Management Centre (DMC); the National Building Research Organisation (NBRO); the National Disaster Relief Services Centre (NDRS); and the Department of Meteorology (DoM), who play a specific role within the phases of the disaster cycle, including early warning, preparedness, relief, response, and/or recovery. On an implementation level, the Ministry Secretary is the Chair for the National Disaster Management

Coordinating Committee (NDMCC), established for coordination of disaster-related activities between all relevant agencies (MNPEA and MDM, 2017).

Following the NCDM, the DMC was founded to implement NCDM directives. More specifically, from an administrative viewpoint the DMA 2005 makes provision for the DMC who are the main government organisation responsible for 'implementation': the coordination and operation of disaster-related activities (Malalgoda et al., 2016; MDM, 2019d). Therefore, the roles of DMC are varied but generally relate to the development of policy, coordinating, monitoring and implementing activities as associated with the National Disaster Management Plan and National Emergency Operation Plan. The DMC also plays a vital role in education and training, as well as to coordinate and monitor risk assessment, disaster mitigation and preparedness activities. In addition, they focus on leadership planning through disaster risk assessment and research, forecasting and the issue of early warning notifications (MDM, 2019d).

An important role of the DMC is as the lead organisation responsible for the issue of early warnings related to disaster situations (Malalgoda et al., 2016). Information dissemination is often via the District Disaster Management and Coordination Units (DDMCU), established under the District Secretariat divisions. A range of technologies and media sources developed to communicate warning messages and include wireless walkie-talkies, SMS and email notifications, and notification via social media platforms. District warnings are then issued on a local level to communities. Further sources of information dissemination are the police and armed forces, generally via public address systems or loudspeakers, given their role at an operational level with rescue and relief activities (Maund et al., 2018; MDM, 2019d).

The NBRO maintains a primary focus on promoting disaster mitigation, preparedness and safety through education, research and training. In this respect, the NBRO offers assistance with, for example, geotechnical and foundation engineering, landslide and slope stability research and disaster mitigation research. Furthermore, they provide specialist services associated with development activities: testing of construction materials, environmental monitoring and assessment. Within the NBRO, building research and landslide risk management form two primary operations. The NBRO also operates as the technical arm of the MDM. Subsequently, during times of emergency, the NBRO contributes to early warning processes: those to related organisations and the media. Following a disaster event, the NBRO employs its technical expertise around risk assessment, including identification of areas potentially vulnerable to secondary impacts following a disaster event. Subsequently, the role encapsulates damage and loss assessment activities, given their knowledge and understanding of regions and impacts (Maund et al., 2018; MDM, 2019e). Following which they assist with resettlement activities and Community Based Disaster Risk Management (CBDRM) programmes (MDM, 2019e).

The NDRS provides a different perspective, with priority given to preparation, response and recovery operations. Importantly, the service provides immediate response assistance to those affected by a disaster as they are the predominant organisation charged with planning and implementation of relief, rehabilitation and reconstruction activities. The NDRS role is broad as it encompasses pre-event activities, such as education and awareness programmes around hazards, risks and vulnerability, flood-prone mapping to Divisional Level Disaster Response, and Action Plans and disaster risk reduction programmes. During events, their role involves, *inter alia*, search and rescue operations and establishing disaster resource centres, covering activities associated with assisting victims. The NDRS role further extends to post-events in terms of monitoring, coordinating and evaluating response and recovery activities (MDM, 2019f). Additionally, their role includes the management and distribution of foreign aid to ensure orderly processes around donated equipment and supplies (Maund et al., 2018).

Further associated with the MDM hierarchy is the Department of Meteorology (DoM) whose role involves, for example, providing meteorological and climatological information and early warning systems with regard to hazard events (MDM, 2019g). The primary role of the DoM concerns the issue of early warnings. In many cases, warning focuses upon forecasts: weather predictions based on scientific and technical evidence. However, at times this may eventuate as special statements to advise of potentially dangerous conditions. During such forecasted events, the DoM is involved with convening meetings between key stakeholder organisations such as the DMC and NBRO to advise of potential climatic conditions and provide early warning and subsequent actions (MDM, 2019g; MNPEA and MDM, 2017).

Although the above discussion has focused around the primary government organisations involved with disaster-related activities, it is acknowledged that there are many government and non-government (NGO) agencies who provide further support during disaster events, providing ground-level emergency assistance. For example, the Department of Irrigation, who provide assistance with early warning data, as well as the Police and Armed Forces for rescue and relief operations (see Figure 6.6).

Alongside the numerous agencies working across the domain of disaster resilience is a plethora of policy to help guide and direct activities. As formerly noted, the Disaster Management Act No. 13 of 2005 (Sri Lanka) governs activities related to disasters including the establishment and operation of the National Disaster Management Plan, the National Emergency Operation Plans, the Framework for Disaster Risk Reduction and the National Emergency Response Plan. However, one of the most significant policy changes for Sri Lanka in recent times concerned Volume II of the Road Map, 'Towards a Safer Sri Lanka 2005–2015' (MDM, 2005; Ministry of Disaster Management and Human Rights 2016.) and the 'Sri Lanka Comprehensive Disaster Management Program (2014–2018)' (MDM, 2014). Both policies are national Ministerial programmes to guide disaster resilience activity.

To implement the DMA provisions, the 'Road Map for Disaster Risk Management – Towards a Safer Sri Lanka' was formulated through the MDM and DMC. This ten-year plan, developed by the Government of Sri Lanka, is a programme of action for disaster risk management (MDM, 2005). Supported by the United

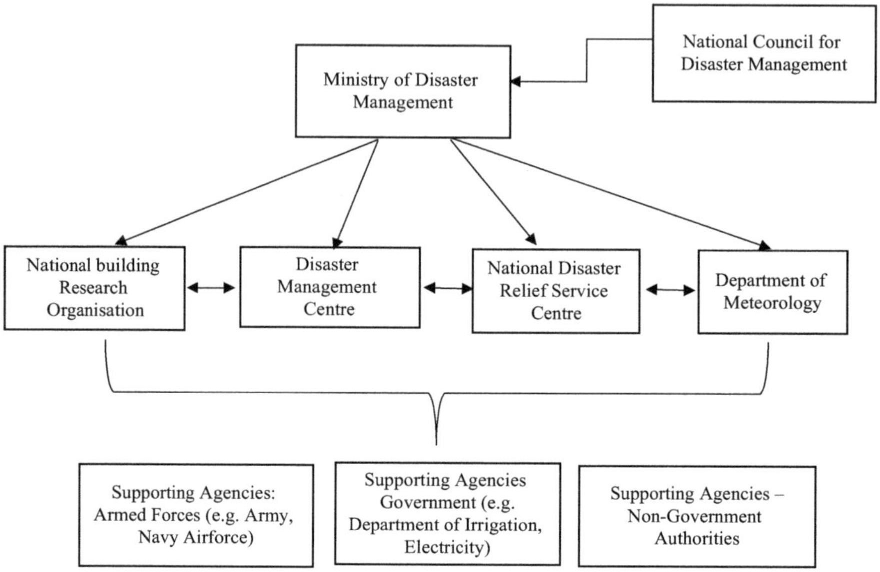

Figure 6.6 Government hierarchy for disaster-related activities (adapted from MDM, 2019h)

Nations Development Programmes, the Plan consists of seven thematic components aligned with disaster risk management activities across Sri Lanka:

> Policy; institutional mandates & institutional development; Hazard, Vulnerability & Risk Assessment; Multi-Hazard Early Warning Systems; Disaster Preparedness Planning & Response; Disaster Mitigation & Integration into Development Planning; Community Based Disaster Management; and Training Education and Public Awareness.
>
> (MDM, 2005, p. vii)

Essentially, the Plan provides a framework of action concerning policy and institutional development with a primary focus on the areas of hazard, vulnerability and risk assessment across multiple levels as identified above. Furthermore, it accommodates actions associated with early warning systems, mitigation measures and importantly evokes changes into existing development planning through integration of disaster risk reduction. Volume II specifically contains project proposals aligned with each of the overarching seven thematic areas. In effect, they are the plans of action including minute details from objectives and budgets through to agency involvement and time frames (MDM, 2005; MDMHR, 2016).

The more contemporary Plan: Sri Lanka Comprehensive Disaster Management Program (2014–2018) (SLCDMP) has been formulated to ensure the 'safety of Sri Lanka by reducing the direct and associated potential risk of the country and minimising impacts on people, properties and the economy.' (MDM, 2014, p. xii). With a focus on disaster-related programmes, it facilitates collaboration with regards to multi-hazard events and importantly aligned with the 'Hyogo Framework for Action 2005–2015' (MDM, 2014).

The SLCDMP presents seven primary objectives that focus on issues such as building capacity at both individual and organisational levels, the integration of disaster risk into development, measures by which to prevent or mitigate disaster impacts, improve coordination and improve response capacity, integration of monitoring and evaluation processes, and increasing disaster risk reduction knowledge (MDM, 2014, p. 17).

It is worth noting at this point that following a major disaster event, under the leadership of the Ministry of National Policies and Economic Affairs (MNPEA) and the Ministry of Disaster Management, a rapid post-disaster needs assessment is undertaken. The report, generally in collaboration with international support through the United Nations, the World Bank and the European Union, assists in the development of a recovery framework and a plan of action. Ultimately the intent of the plan is to address short, medium and long-term recovery requirements and also focus on future building activities. Subsequently, the plan enables economic evaluation to support recovery actions (MNPEA and MDM, 2017).

An additional area is related to policy concerns funding to assist those affected. The Government of Sri Lanka supports recovery of local communities primarily through two policy instruments that present a budget of approximately US$168 million: the National Insurance Scheme (NIS) and the Catastrophe Deferred Drawdown Option (CAT-DDO). The NIS has a primary focus on lives and property, including homes and businesses impacted by the disaster event. While the CAT-DDO presents a different arrangement and, although aligned with relief and recovery, it is more nationally focused. Under this scheme, the World Bank provides loan access for up to US$102 million (MNPEA and MDM, 2017).

The above discussion provides an insight into the disaster resilience policy in effect across Sri Lanka at a government level. The Government maintains central organisations that oversee policy and initiatives to assist in preparedness, rescue and relief, as well as with recovery operations. However, with major hazard events, local communities continue to experience devastation with loss of life, destruction of property and disruption to services and livelihood that all have a major economic impact on those affected and on the country on a wider scale. Therefore, the intent of this research was to further explore the policy implementation process to identify the challenges associated with disaster resilience policy and understand areas that require change.

Understanding the 2017 flooding and landslide events

The former discussions have provided an insight into the rich nature and history of the country of Sri Lanka: a beautifully isolated island, yet one at the mercy of devastating hydro-meteorological hazards. The 2017 flooding and landslide disaster events made a significant and notably negative contribution to the landscape of the country, as they accentuated the vulnerabilities of local communities. For countries such as Sri Lanka that proactively encourage disaster risk resilience activities, it is paramount to be able to identify challenges experienced across disaster resilience operations and seek mechanisms by which to reduce vulnerability and increase resilience.

The 2017 disaster events produced significant devastation to many regions across Sri Lanka. Therefore, the intent of this section is to first provide an overview of the range of impacts following disaster events on a national scale to provide context in terms of the devastation. Following which the focus will turn to the case study site: Ratnapura and the local level experiences of those responsible for policy implementation.

The devastation of the 2017 events resulted in flooding and landslides affecting more than 15 Districts, with five predominant epicentres of activity, impacting more than 800,000 people. As formerly detailed, more than 200 people sadly perished during the event, with others injured and many who remain missing. With such devastation, it has been estimated that damages to the five significantly affected districts equates to a staggering LKR 70 billion (around US$385 million) (MNPEA and MDM, 2017).

As with many hydro-meteorological disaster events, housing, agriculture, transport and industry sectors are negatively impacted – this was the case in Sri Lanka. The violence of the disasters resulted in significant damage to health centres, educational facilities as well as water and sanitation services. The aggressive nature of flooding and landslides, in conjunction with their subsequent flow-on effects, brought many regions of the country to a standstill as the devastating impacts came to be realised. In particular, the District of Ratnapura has been identified as one of the worst affected areas following the 2017 flood and landslide disaster events. The number of reported deaths for the region was significant, totalling 86. The region faced true devastation with significant flooding, and suffered more than 18 of the 35 major landslides and multiple minor ones (MNPEA and MDM, 2017).

The evaluation of policy serves to identify challenges experienced as a result of the 2017 events and to identify mechanisms for change and growth, ultimately aiming to reduce vulnerability and foster community resilience. Therefore, the intent of this section is to explore those challenges from a policy perspective – including implementation from the point of view of those persons engaged in disaster risk resilience and recovery – in order to identify strengths and areas to improve to assist those most vulnerable.

A national overview

The following content related to impacts on a national scale has been garnered primarily from the Ministry of National Policies and Economic Affairs and the Ministry of Disaster Management report, 'Sri Lanka Post Disaster Need Assessment', issued in May 2017.

Organisations including the MNPEA and MDM released a Rapid Post Disaster Needs Assessment following the 2017 event that detailed the effects associated with life and property. According to the MNPEA and MDM report, the impacts of floods and landslides across districts were devastating, with significant loss of life, injury and persons missing, as shown in Table 6.2.

Table 6.2 Impacts of floods in four highly affected districts (source: MNPEA and MDM, 2017, p. 21)

	Matara	*Ratnapura*	*Kalutara*	*Galle*
No people affected	183,008	235,197	192,464	158,693
Deaths	32	86	70	16
Injured	16	99	20	4
Missing	–	15	43	15

The MNPEA and MDM report provides a range of economic implications following the 2017 event; a number of these figures are highlighted in this section, as they provide the context for understanding the severity of disaster in terms of both impact and economic consequences.

From an infrastructure perspective, the devastation of the flooding and landslides had a profound effect on health services across multiple districts. Damages were wide-ranging, from basic infrastructure services and equipment, through to loss associated with additional expenditure for emergency response and medical supplies. Across Kalutara, Ratnapura, Galle and Matara alone, the MNPEA and MDM (2017) report that 65 health care institutes were affected, involving hospitals and medical offices. Important to note is that a number of these institutes had, unfortunately, been designed and built in high risk areas. Again, Ratnapura was considered a district with significant levels of damage.

The housing sector presented a similar story, with damage across multiple Districts – many were completely or 'fully' damaged to the extent they were rendered inhabitable (refer Figure 6.7). Although many Districts were affected, it has been highlighted that more than 80 per cent of the housing impacts were experienced by those residing within rural areas. The MNPEA and MDM (2017) state the damage to the housing sector to be around LKR 29 billion (around US$160 million) clearly identifying the devastation experienced by local communities.

As previously discussed, many urban and agricultural practices have reshaped the natural environment, resulting in destabilisation of local lands and modification of waterways, which contributes to vulnerability during hazard events. In addition, the practice of sand mining is another factor that created unstable river banks, as bank vegetation became uprooted by flood waters and subsequently blocked natural river pathways. Cultivation of land for crops such as tea and the uprooting of rubber plantations provide two additional examples of change

Figure 6.7 A house destroyed by a landslide in Ratnapura district in 2017

that continues to influence landslides, given the degree of instability brought about by such practices.

Economically, the impacts of the 2017 event were widespread and affected multiple sectors. Notably, for the 15 disaster-affected areas, many were acknowledged as significant contributors to the agriculture, trade and service sectors. Although the fallout following such a disaster event may be estimated, it is expected that damages, losses and needs may far exceed those estimations, given flow-on effects to livelihoods, businesses and service providers. Agricultural practices, from plantations and livestock sectors, were deeply impacted by the 2017 events, with many land areas remaining submerged for days. Plantations, particularly in low-lying areas, suffered immense damage – the MNPEA and MDM estimate that over 3,000 hectares of land and buildings used for tea plantations were damaged or totally destroyed.

Both formal businesses and those associated with home-based livelihoods were affected, impacting hundreds of thousands of people. On a local level, the damage to lands impacted many community members whose livelihood was dependent upon home occupations based upon home agricultural produce. Given the degree of impact on both formal and informal businesses, this produced a negative effect upon local labour forces. With a high degree of land flooding, the agricultural sector was most severely affected, particularly given that many regions were made inaccessible. Therefore, with significant devastation across lands, existing labour sources ceased for extended periods.

In terms of major infrastructure, those associated with irrigation and bridge structures were damaged across multiple districts. In some cases, elements of irrigation and bridges became dislodged and combined with loose vegetation to present a further hazard as they contributed to the formation of blockages. During the event, many areas received continuous torrential downpours, with the Kalu Ganga Basin Station recording rainfall of 615mm on the day of the main disaster, while Ratnapura and Kalutara Putupaula hydro stations recorded 326mm and 296mm respectively. The rain intensity and the subsequent fast-flowing flood waters resulted in significant damage to many flood protection schemes that had been designed to protect communities. Almost unbelievably, the Nilwala River overflowed the spillway at the Panadugama bypass by a staggering 8.7 metres. Ultimately, the continuous rains combined with the force of the flood waters resulted in many spillways becoming blocked with vegetation and storm debris; this resulted in upstream flooding and severe damage to bunds, rendering them ineffective.

Critical road infrastructure faced similar issues, as flooding, subsequent landslides and the risk of landslides affected most transportation services. The majority of roads were impacted to some degree across the five major districts affected, with nearly 900 road sections sustaining carriageway damage and many more having bridge, culvert, drain and erosion damage. On a national level, the MNPEA and MDM identify that 55 road sections were significantly damaged; a similar situation was presented for provincial roads, with around 89 sections presenting with substantial damage. In many cases the road networks were completely unpassable until flood waters subsided. Even then, many roads and bridges were unable to support regular transportation services. Therefore, the damage to road services had a significant impact on transportation and ultimately on many business communities.

The above discussion from MNPEA and MDM (2017) provides a clear illustration of the degree of devastation following the 2017 flooding and landslide events. The adverse impacts remain undeniable; however, an understanding of the policy implementation process is needed to explore areas that could be strengthened to assist in preventing such future devastation when the next disaster event strikes.

Ratnapura: challenges of disaster resilience

The following is sourced from Maund et al. (2018).

Ratnapura, a large town located on the Kalu River, is situated south-east of the national capital Colombo. In line with other parts of the country, Ratnupura is subjected to extreme hydro-meteorological events, including severe floods and landslides (Dilhani and Jayaweera, 2016). Unfortunately, human activities have exacerbated extreme weather events across this region (UN-Habitat, 2013), as communities reshape lands through urbanisation and agricultural practices.

In a similar situation to its neighbouring towns, lands cultivated for the purpose of crop plantations were damaged and crops destroyed in the 2017 Sri Lanka floods, as flash flooding forced its way across land parcels and landslides occurred. In these areas, changes to land structures to accommodate infrastructure services such as roads and water systems have effectively reshaped land to produce a combination of steep slopes and diversion channels that may then impede or halt natural drainage pathways. Waste management presents a further major challenge as it often encourages flooding, as land disposal methods have physically altered pathways used to naturally disseminate excess water.

As mentioned in the previous section, the 2017 floods led to significant loss of life, property destruction, loss of livelihood and continual economic consequences. Understanding the key challenges facing those responsible for disaster-related activities is an important step towards future policy formulation and implementation. The research conducted as part of this study adopted a two-staged qualitative exploratory research design. First, academic and grey literature was analysed, including media articles from six national newspapers in Sinhalese, English and Tamil across a four-month period from the May event, along with international media reports. Following this, semi-structured interviews were conducted with key stakeholders from both the government and non-government sectors. In most cases interviewees were senior professionals whose roles were aligned with policy formulation, implementation or response related to disasters.

In terms of challenges in disaster resilience policy implementation, a range of operational areas were highlighted, including technology, operational and departmental structural issues. The challenges are now presented with consideration of the major issues identified. It is important to note that many challenges are interrelated; however, individually they provide a more thorough representation.

Key considerations

A range of challenges were raised following the 2017 floods, particularly in relation to Ratnapura. It is interesting to note that the frequent occurrence of disasters was identified as problematic to effective disaster resilience and management operations. Given the geographical isolation of Sri Lanka, it regularly experienced a host of extreme disaster events – these are becoming more and more frequent. Thus, the government has to respond quickly and continually to a range of different situations. Given the short time frames between the next disaster, this presents a constant pressure in terms of responding to events and attempting to provide relief assistance and recovery work. Although multiple challenges were identified, there were three that were of particular concern among most stakeholders and were often raised during interviews and focus group sessions. These were: early warning systems; emergency response and relief operations; and mitigation projects.

In terms of early warning systems, there were two main issues highlighted: weather forecasting and communication. The conditions across the country vary rapidly due to its location near the equator, combined with its isolation as an island and geographical characteristics such as mountainous regions, therefore forecasting itself was identified as problematic. In this respect, technology presented a barrier to effective early warning communications. Technology for forecasting disaster events and to enhance early warning activities – such as automated weather station and surveillance radars – were generally unavailable within the DOM. Therefore, in relation to the 2017 floods, timely and accurate forecasts and the provision of early warning communications were considered an issue; this was a direct reflection of the lack of technology available. Observations of wind data and forecasting numerical modelling were conducted at the time; however, there was a need for computers with extensive processing powers to enable true forecasting capabilities.

In general, the weather forecasting systems employed at the time often related to larger regions or areas – subsequently, local rainfall intensity data was minimal. In effect, this means that communities were often not prepared for an event, as forecasts and media communication for a larger region was unable to garner significant local attention to assist with early warning and evacuation activities. Implementation of local area forecasts was considered vital for early warning: it was explained that 100mm of rainfall over one hour may have disastrous consequences, whereas the same amount of rainfall over a 24-hour period may have minimal impact. Therefore, relaying this information on a local level was identified as vital for early and rapid evacuation of those high-risk areas (refer Figure 6.8). Following the 2017 event, it was identified that the communication system itself failed due to the extreme weather conditions; this was primarily a result of

Figure 6.8 The Department of Agriculture in Ratnapura provides flood forecasts by radio based on flood level gauges, but the extent to which they are followed by communities is uncertain

network breakdowns or electricity disconnections. Therefore, a review of the existing communication system and the introduction of measures to ensure the protection and continual operation during disaster events was viewed as a crucial resilience measure. Other means of communication were employed; however, for those people residing in the more remote areas of Ratnapura, there was significant difficulty in disseminating information.

From an emergency response and relief perspective, there were major challenges related to supplies, equipment and organisation of local level operations. The ability to source supplies and equipment and provide to those in need presented a challenge, given the ferocity with which the disaster hit and also the high number of people impacted across Ratnapura.

Limited availability of response and relief equipment presented a significant challenge, as operations were often hindered by insufficient supplies. Basic essential items were often lacking, such as jackets, tents, towels and matches, while vital supplies such as bottled drinking water were often unavailable, presenting a significant challenge. It was noted that at one stage a small number of bed sheets were available to evacuees, yet those evacuated numbered in the thousands at that particular time. Similarly, for three days preceding the event, some regions faced a complete lack of certain supplies such as mats. On a broader scale, it was seen that evacuation authorities lacked large-scale equipment such as boats for rescue operations during flooding. With transport to many affected areas hindered due to inaccessible roads and bridges, this resulted in significant delays in response and relief activities; as a result, assistance was sought from the Navy.

The distribution of supplies often presented a difficult situation to organisations, given that many people resided outside main affected areas. Without knowledge of exact locations, there were delays and difficulties in reaching those in need of assistance. It was highlighted that due to the involvement of multiple organisations for distribution of supplies, lack of communication often resulted in some areas being over-serviced while others continued to remain in desperate need of support. As discussed previously, the ability to provide emergency rescue and relief activities was often hampered; this issue may provide an explanation for why community members were hesitant to evacuate their homes when notification was provided.

In terms of disaster resilience, mitigation projects often present a range of challenges, particularly as illegal construction in the river reservations tended to encourage the negative impacts of disaster events. Although the government has the legal authority to remove such constructions, the legal process of litigation was identified as both time consuming and costly. In addition, insufficient human resources and the political environment often rendered this form of action unrealistic; therefore, illegal construction continues to occur.

For Ratnapura specifically, the town area has a history of major flooding events – in the 1990s it was actually relocated to higher grounds in an attempt to mitigate future disaster impacts. However, gem mining remains a major form of livelihood for many local communities, and is a major tourism feature; for this reason, abandoning the original town has not been considered by many. From a disaster resilience and mitigation perspective, apart from the issue of relocation, the exact location of mining tunnels remains unclear and as such hinders development of both scientific and disaster risk planning activities. Further relocation of community members as a means of reducing risk has been considered; however, removal from high risk areas is generally impossible given the scarcity of land available for resettlement. In this respect, the focus is moving towards living with disasters: to resilience and mitigation operations. This may involve specific development activities, potentially adherence to safer building codes and urban planning, through to education and awareness, in addition to improved early warning systems.

The local level experience

Across the interviews and focus group sessions, a range of additional challenges were highlighted; these will now be explored, as they provide insight across multiple levels that can be used to inform policy and building resilience. Many of the challenges described here are closely aligned with those discussed in the former section; however, they present a range of additional issues that require consideration to build future capacity and resilience on a local community level.

Community preparedness, or lack thereof, was highlighted as a major challenge to those responsible for relief activities. Although community members were often aware of and provided with notice of an impending event, there remained a lack of preparedness – particularly in terms of readiness to evacuate homes. Although some community members were considered to be aware of impending events and subsequent consequences of remaining in their homes, there was a level of unwillingness to self-evacuate. Many communities delayed evacuation until there remained no other choice. Ultimately, this placed significant pressure on agencies involved with disaster risk response activities, as their focus was often directed towards evacuation. If communities relocated during early warning notification periods it was viewed that more attention could be focused on relief activities and the provisions to support those who had been displaced.

Traditionally, when a disaster event is imminent, local communities shelter in the nearest temple, school or similar building. However, as there was a lack of formally dedicated safe shelter points during the 2017 floods, this meant residents were dispersed randomly amongst regions, potentially within hazard paths. Although many buildings were eventually nominated for emergency shelter purposes – safety havens for many community members seeking emergency refuge or for those who had become displaced – it was noted that in most cases these 'relief' structures were not designed to accommodate basic needs and serve as safety shelters. Thus, this led to a range of issues associated with water, sanitation, safety, gender and privacy. Designed safe shelters were seen as important to fostering a sense of preparedness, as they can provide direction for community response during emergency events. Additionally, they may provide the government with a clear understanding of locations in which residents are sheltering to assist with relief operations, and where necessary, future evacuation activities.

Lack of coordination across the disaster relief process was viewed as problematic: both internal and external across government and non-government agencies. It was clearly identified that there remained a delayed response in activating services. Furthermore, with multiple agencies supporting relief actions, duplication and wastage was highlighted as a major concern, as formerly discussed. There appeared to be no overriding coordinating body to manage relief items, and as such there remained no obvious coordination or collaboration between the various agencies involved with disaster relief activities. Interestingly, a number of the non-government agencies providing support were not considered to have any legal role or policy mandate to undertake such activities, and therefore were viewed as competing with other agencies; it seemed that they had a marketing interest in participating and in this vein often hindered coordination attempts.

Lack of coordination was strongly viewed as a result of bureaucratic process. As with many countries, government operations within Sri Lanka often reflect a highly bureaucratic society and in turn this has the potential to result in lengthy administrative approvals or response actions that are both unsuitable and inappropriate during an emergency situation. Unfortunately, during the 2017 floods, there was a delayed governmental response to the disaster across many areas due to bureaucratic operations. Even when planning was rapid, the actual ability to enforce operations was viewed as hindered and delayed.

Delays were further exasperated by officials not wanting to work outside of standard operating procedures, even though it was an emergency situation. Many government agencies were involved with disaster-related activities and in most cases all had specific agendas, policies and procedures to adhere to. Although a need for adherence to procedures is acknowledged, there needs to be flexibility in operation to accommodate unforeseen emergency events. In the case of an unusual event, any change to procedure may have a profound impacting response time. A typical case from a previous event involved a substation rainfall monitoring station, where an officer identified that the fall was unusually heavy and reported the occurrence that provided early warning; however, this was viewed as unusual to report such a circumstance and outside of standard procedure.

From a planning perspective, concern was raised over mapping, risk assessment and evacuation drills, particularly in terms of unavailability of maps. Disaster response activities often involve use of mapping related to specific hazards to identify risk areas or those likely to be affected by a pending event. During the 2017 event, mapping that related to landslide events was accessible; however, there remained no appropriate flood mapping available to assist with response activities. There was a general consensus that there was an opportunity to develop mapping, given the 2017 flood events, in conjunction with satellite images and site visits. However, at that time there was a dearth of appropriate flood mapping; this may be a result of the need for additional human resources and specialisations in flood map preparation.

In the same way, disaster risk assessment was also viewed as a factor affecting disaster response. Risk assessment was identified as a major challenge facing Sri Lanka, with the current governmental hierarchy and policy directions that often negate inclusion of these activities on a local level. Areas such as hazard assessment and hazard profile combined with vulnerability profiles were deemed to be necessary to identify those most at risk, assist with preparedness and build capacity and resilience. From a resilience perspective, inadequate regular evacuation drills were also seen to contribute to a lack of community preparedness. Across the country, evacuation drills are conducted for tsunami events to help raise awareness and prepare communities. Although flooding and landslides are frequent events facing many regions – and are becoming more frequent – at present there remains no evacuation drill process in place nor is it mandated by policy. For a range of reasons, communities were seen to have become complacent in terms of preparing for disaster events; where exposure to an event was not experienced within a specific time period, indifference and a loss of interest appeared to set in.

Post-disaster challenges were also identified and it was acknowledged that the lack of proper guidelines for reconstruction affected future decision making – regulatory policy, or even guidelines, were unavailable to direct reconstruction activities. Recovery is addressed through policy; however, areas related to reconstruction activities lack direction. Reconstruction is an important post-disaster consideration to enable sensitive and appropriate design to assist with building resilience against future events. At present, there is limited ability to guide reconstruction activities, and although policy was highlighted this was also seen to align with a need for technical expertise.

Resettlement opportunities

Following the 2017 floods, resettlement was identified as a significant challenge – particularly for those residing within Ratnapura, as well as on a national level. Resettlement presents an opportunity to minimise future disaster risk. Yet the nature of the process presented much complexity for those involved with implementation, particularly in the absence of policy.

Although there was evidence of a resettlement policy for internally displaced people (for example, related to conflict) there appears to be no overarching formal resettlement policy related to disaster resilience. Although it was identified that the NBRO has a resettlement procedure, development into a formal policy was highlighted as necessary at a higher government level. One of the major concerns related to land owners returning to affected lands and again settling in places that put them at a high risk of future events. It is worth noting that following major tsunami events, information policy documents were developed that focused on resettlement; however, given their nature towards one specific disaster, their adoption on a national level did not eventuate.

In a similar vein, an absence of an authority to oversee resettlement activities was acknowledged as a challenge. Nationally, Sri Lanka has a structured governmental hierarchy for disaster risk reduction and management. However, following the main disaster event and subsequent relief processes, there appears to be a gap in terms of a dedicated authority to manage resettlement. It has been identified that each District Secretary tends to oversee such activities; however, they are responsible for large areas and given existing administrative functions there would potentially be a conflict amongst priorities. Additionally, given their existing administrative functions, there is a need for further resources and professionals to manage resettlement, yet in the current context this remains absent.

On more of a local level, Sri Lanka is rather small in terms of land area and, in this respect, the availability of land suitable for redevelopment remains limited. Given the hydro-meteorological events that frequent the country, the ability to locate land that is effectively 'disaster free' is somewhat difficult, if not impossible. Therefore, risk assessment is often a mechanism employed to assist with identification of suitable lands for resettlement of people displaced due to disaster events. Unfortunately, this means resettlement in regions prone to future hazards. The livelihood of local communities is often connected to the areas or regions in which they reside: many people in rural areas depend on agriculture as they often cultivate the land on which they live. Although they may face a loss of livelihood following a flood event, they may be hesitant to relocate, presenting a challenge for policy makers.

Across community, government and non-government agencies, the same issues arose with regard to a lack of coordination amongst stakeholders. In this respect, resettlement activities were seen as problematic and challenging, particularly between government and non-government organisations. Potentially, the absence of a dedicated authority for management of resettlement, as discussed above, contributed to existing coordination challenges. It was explained that an NGO identified a transitional shelter and made significant land changes. Communication with the NBRO would have identified this land parcel as an ideal location for a permanent shelter, which was no longer possible after changes to the land. There have also been instances where the government implemented plans to incorporate resilient features; however, these are not always known or have no mandatory policy mechanism for enforcement. Subsequently, there have been instances where a government agency directed an NGO to follow their own design, demonstrating a lack of communication and coordination as well as a dearth of understanding of the importance of objectives with risk mitigation.

These examples highlight the importance of effective communication and collaboration: to use disaster events as an opportunity to redevelop rather than intermediate solutions. It was highlighted that the National Physical Plan presents the opportunity to resettle people in specific urban areas where planning and design have directed facilities, services and livelihood options. Through a coordinated programme, resettlement has the potential to be effectively implemented.

Key strategies of success

Although it is important to identify key barriers and challenges to successful policy implementation for future change, it remains just as important to consider areas of strength, particularly those areas addressed following extreme weather events. Through the identification of successful strategies, 'best practice' can be formulated and positive attributes shared across other project areas. Following the 2017 Sri Lanka floods, a number of key strategies were identified that have contributed to early warning and evacuation, ultimately resulting in preservation of lives. Although these strategies were considered successful, they are yet to be fully implemented on either the local or national level, and they focus on one-off incidents or are currently in process following the 2017 events.

Education programmes aimed at school students have shown potential for improving capacity building through awareness and preparedness. In one particular situation, such awareness directly saved the lives of school students who had undertaken a landslide education programme. In this programme, students developed an understanding of how to recognise the early signs of a landslide event. Subsequently, during a heavy rain event, students were able to identify the sloping of nearby trees; as a result, the teaching staff decided to immediately evacuate four classes. Within a short time frame, the entire building collapsed. Thus this example shows how educational initiatives at a grassroots level can result in increased levels of awareness that may save lives.

In light of former discussions, community-based early warning systems were considered to be effective mechanisms to assist with evacuation activities. Even very basic early warning systems have had an impact on life safety through rapid evacuation. Distribution of plastic rain gauges to community members prior to the 2017 floods, combined with education and training on rainfall levels associated with landslide risks has directly assisted in early evacuation and saving lives. Communities who had been engaged in rainfall education managed to evacuate at critical moments, while many who ignored warnings sadly lost their lives. Knowledge and empowerment have provided locals with the opportunity to measure rainfall and evacuate when rainfall reaches warning levels, resulting in rapid evacuation without reliance on government intervention. However, the question now becomes how to engage all community members in this awareness campaign.

Within the DMC, the establishment of the National Emergency Operation Centre (NEOC) to allow quick decision making during emergencies. The NEOC is a unit that consisted of personnel from each of the major government organisations involved with disaster rescue and relief operations. During times of emergency, established bureaucratic processes often delays decision-making activities. However, through the NEOC unit having representatives from relevant government agencies, in many instances enabled effective and rapid decision making. A further successful strategy by the DMC has been registering the suppliers of emergency relief. The intent is to provide a standard operating procedure to ensure emergency relief items are available and distributed in an effective and timely manner. In the form of a Memorandum of Understanding between the DMC and suppliers, it was agreed that food packets must be supplied within the first six hours of an event. In effect, this strategy assists with coordination of relief operations amongst organisations and across regions.

The police and armed forces played a vital emergency response role in the 2017 floods. Through their intervention, communication between various agencies and the local community during emergency situations was enhanced. Similarly, air force personnel used helicopters to assist with rescuing people stranded in inaccessible areas. Naval officers also used their vessels for rescue operations and relief distribution. Collaboration between government, NGOs and police and armed forces in evacuation and relief activities has been considered a successful strategy for inclusion in future planning.

As formerly discussed, historically there have been no designated evacuation centres and communities impacted by disaster events have sought refuge in often ill-equipped buildings. These random and makeshift evacuation centres have the potential to hinder evacuation and relief operations, as formerly discussed. However, the government have officially established three designated evacuation centres in high-risk areas. The formalisation of these three centres has been well-received across the regions and it is understood that further centres will be established that also address many of the former concerns raised over building suitability.

Multiple long-term mitigation projects associated with flood and landslide events have been established by the government. The Mahaweli project provides a typical example, with a major outcome being the control of flooding in downstream areas, while the Colombo project has involved the development of multiple inland reservoirs to increase retention capacity. The multitude of projects have involved works around stabilisation, water pumping and similar activities that all focus on long-term mitigation works to reduce risk and impact.

Capacity building has the ability to reduce disaster risk, particularly when focused at the community level; planning and building form a significant factor in this area. Implementation of safer building codes has often presented a challenge due to insufficient local level design requirements and potentially policy enforcement capabilities. The NBRO has taken a proactive approach to build capacity of those involved in construction activities through a targeted training programme. Training has involved around 5,000 masons so far, with guidelines developed to assist in understanding and complying with building codes, thus assisting with structural stability and fire safety measures, and it is understood that this played a significant role in allowing time for evacuation.

A changing policy environment

In the preceding sections, a number of key challenges have been identified in relation to disaster resilience policy implementation following the 2017 Sri Lanka floods. It is important to note that formal policy reviews are currently under way and multiple policy-related actions in process to avoid such future devastation through disaster resilience.

One major policy action involves the formal establishment of a new Flood Management Act. Although policy currently exists, it was deemed that a new Act was required to adequately cover current climate-related events and represent contemporary disaster resilience practice. In addition, policy related to evacuation is currently under consideration following the reluctance of some people to evacuate. Even after early warning notices were issued and people in locations of risk were encouraged to evacuate, many refused to do so – this ultimately placed them at significant risk. In many cases, failure to evacuate prior to critical events meant that when the disaster struck, it was too late to leave. Therefore, evacuation is an area that is being considered at a government policy level to identify options to force evacuation in an attempt to save lives in future.

As highlighted, there is a definite need for designed shelters to be formally established as places of safety during times of emergency and for those who have become displaced. Although a number of shelters have been established, there is now a focus on formalising further buildings in this capacity, with an additional 100 buildings designated for this purpose nation-wide. As designated safe havens, they can be selected to ensure appropriate accommodation with regards to issues of privacy, security and the like.

In terms of policy change, one of the most major areas currently under investigation relates to the opportunities for relocation of people currently residing in areas identified as high risk. The intent is to minimise deaths and property destruction from disaster events and the subsequent hardships associated with economic and infrastructure impacts. The complexity and challenges of such a proposal are well acknowledged; however, the provision of alternative lands to those residing in high risk areas is considered a positive step in mitigating disaster impacts.

Conclusion

Sri Lanka is a country that often experiences hazards, particularly those of a hydro-meteorological nature, with events becoming more and more frequent. The country faces a significant set of challenges due to its geographical location and exposure to climate change impacts. However, disaster events are often exacerbated by urbanisation and agricultural practices which contribute to destabilisation of land and decrease the ability of natural waterways to effectively and naturally disperse high volumes of water during floods.

The 2017 floods and subsequent landslides have had a profound affect on the country and show a picture of true vulnerability and devastation: significant loss of life, property destruction and ensuing economic consequences on a local and regional scale. Sadly, as the death toll and number of people adversely affected by this event shows, there is a desperate need to further build capacity and resilience in Sri Lanka.

In this respect, the Government of Sri Lanka has been proactive in implementing a range of strategies to direct disaster risk reduction and response activities, including the establishment of the DMC and the Sri Lanka Disaster Management

Act in 2005. The Road Map, 'Towards a Safer Sri Lanka 2005–2015' and the 'Sri Lanka Comprehensive Disaster Management Program (2014–2018)' indeed demonstrate true commitment through action. Subsequently, a range of successful strategies have evolved around community capacity building, early warning systems, government intervention in evacuation and relief operations, as well as long-term mitigation schemes.

However, the 2017 disaster event highlights the need for yet further intervention, given the devastation experienced across the country, highlighting vulnerability to future events. Primarily, the research has shown a need to continue to develop policy in the following areas: community preparedness activities; improved organisational communication and coordination; early warning systems; policy flexibility under emergency situations; reduced bureaucratic processes; and importantly, planning and building enforcement.

From a Government perspective, a number of these issues are being addressed through the establishment of a new modernised and relevant Flood Management Act. Additionally, given the difficulties convincing those at risk to evacuate, consideration is being given to the development of an evacuation policy. A number of designated safety shelters have been established, with more proposed, and Districts have been provided with supplies relevant to population. Furthermore, investigations into permanent relocation of local residents outside of high risk areas is currently under review.

Although Sri Lanka has numerous agencies and regulatory policies governing disaster-related activities, there remains minimal post-disaster evaluation of policy and operational effectiveness. The intent of this chapter was to explore the area of implementation effectiveness, given the continual hazards experienced by the country and the tragic loss of life and property devastation. In this context, evaluation can enable an understanding of current practice and how to direct future policy change to assist building capacity and resilience on a local level.

References

Ashraf, S., Luqman, M., Iftikhar, M., Ashraf, I. and Hassan, Z. (2017) 'Understanding Flood Risk Management in Asia: Concepts and Challenges' in Hromadka, T. and Rao, P. (eds.) *Flood Risk Management*. IntechOpen. Available at: https://www.intechopen.com/books/flood-risk-management/understanding-flood-risk-management-in-asia-concepts-and-challenges. DOI: 10.5772/intechopen.69139.

Central Bank of Sri Lanka. (2017) *Economic and Social Statistics of Sri Lanka*. Colombo: Central Bank of Sri Lanka.

De Silva, R. and Punyawardena, B.V.R. (2013) 'Drought' in Disaster Management Centre & Ministry of Disaster Management. (eds.) *Hazard Profiles of Sri Lanka*. Colombo: Disaster Management Centre and Ministry of Disaster Management, pp. 046–063.

Department of Meteorology Sri Lanka (DoM). (2019) *Climate of Sri Lanka*. Available at: www.meteo.gov.lk/index.php?option=com_content&view=article&id=94&Itemid=310&lang=en (accessed 2 October 2019).

Dilhani, M. and Jayaweera, N. (2016) 'A Study of Flood Risk Mitigation Strategies in Vernacular Dwellings of Ratnapura, Sri Lanka'. *Built-Environment: Sri Lanka*, 12(1), pp. 1–9.

Disaster Management Act. (2005) (Sri Lanka). Colombo: Government of Sri Lanka.

Disaster Management Centre (DMC). (2017) *Rapid Impact Assessment Report*. Available at: www.disastermin.gov.lk/web/index.php?option=com_content&view=article&id=177%3A2017-flood-rapid-impact-assessment&catid=1%3Alatest-news&lang=en (accessed 3 October 2019).

Disaster Management Centre and Ministry of Disaster Management. (2005) *Towards a Safer Sri Lanka: A Road Map for Disaster Risk Management*. Colombo: Ministry of Disaster Management.

Disaster Management Centre and Ministry of Disaster Management. (2019) Hazard Profiles of Sri Lanka. Colombo: Ministry of Disaster Management. Available at: www.dmc.gov.lk/images/hazard/hazard/Index.html (accessed 3 October 2019).

Foundation for Environment, Climate and Technology (FECT). (2017) *Climate and Disaster Management*. Available at: www.climate.lk/fect/disaster_index18cb.html (accessed 12 June 2019).

Intergovernmental Panel on Climate Change (IPCC). (2007) *Climate Change 2007: Synthesis Report: Contribution of Working Groups I, II and III to the Fourth Assessment Report on the Intergovernmental Panel on Climate Change*. Geneva: IPCC.

Kumarasinghe, N. (2013) 'Lightning' in Disaster Management Centre & Ministry of Disaster Management. (eds.) *Hazard Profiles of Sri Lanka*. Colombo: Disaster Management Centre and Ministry of Disaster Management, pp. 104–143.

Lalitha, R., Fernando, S. and Muthulingan, A. (2015) 'Effectiveness of Administrative Preparedness: A Case Study on Flooding Conditions in Ambagamuwa Korale Division of Nuwara Eliya District in Sri Lanka' in Ha, H., Lalitha, R., Fernando, S. and Mahmood, F. (eds.) *Strategic Disaster Risk Management in Asia*. New Delhi: Springer, pp. 87–98.

Malalgoda, C., Amaratunga, D. and Haigh, R. (2016) 'Overcoming Challenges Faced by Local Governments in Creating a Resilient Built Environment in Cities'. *Disaster Prevention and Management: An International Journal*, 25(5), pp. 628–648.

Maund, K., Ahmed, I., Gajendran, T., Jayawickrama, T., Karunaratne, T. and Sandanayake, Y. (2018) *Understanding the Review of Disaster Management Policies Following Recent Flooding and Landslide Disaster Events in Sri Lanka*. Newcastle: University of Newcastle.

Mawilmada, N., Atapattu, S., Dela, J., Gunawardene, N., Weerasinghe, B., Nandana, M., Bellanawithana, A., Wimalasiri, R. and Kumari, N. (2010) *National Climate Change Adaptation Strategy for Sri Lanka, 2011-2016*. Climate Change Adaptation. Available at: www.climatechange.lk/adaptation/Index.html (accessed 26 October 2019).

Ministry of Disaster Management (MDM). (2014) *Sri Lanka Comprehensive Disaster Management Programme 2014-2018*. Colombo: Ministry of Disaster Management.

Ministry of Disaster Management (MDM). (2019a) Hazard Profile of Sri Lanka. Available at: www.disastermin.gov.lk/web/index.php?option=com_content&view=article&id=57%3Ahazard-profile-of-sri-lanka&catid=73%3Areports&Itemid=70&lang=en (accessed 30 March 2019).

Ministry of Disaster Management (MDM). (2019b) Sri Lanka Disaster Management Act and Policy. Available at: www.disastermin.gov.lk/web/index.php?option=com_content&view=article&id=90&Itemid=83&lang=en (accessed 2 October 2019).

Ministry of Disaster Management (MDM). (2019c) National Council for Disaster Management. Available at: www.disastermin.gov.lk/web/index.php?option=com_content&view=article&id=51&Itemid=64&lang=en (accessed 2 October 2019).

Ministry of Disaster Management (MDM). (2019d) *Disaster Management Centre*. Available at: www.disastermin.gov.lk/web/index.php?option=com_content&view=article&id=54&Itemid=78&lang=en (accessed 2 October 2019).

Ministry of Disaster Management (MDM). (2019e) *National Building Research Organization*. Available at: www.disastermin.gov.lk/web/index.php?option=com_content&view=article&id=55&Itemid=79&lang=en (accessed 2 October 2019).

Ministry of Disaster Management (MDM). (2019f) *National Disaster Relief Services Centre*. Available at: www.disastermin.gov.lk/web/index.php?option=com_content&view=article&id=92&Itemid=150&lang=en (accessed 2 October 2019).

Ministry of Disaster Management (MDM). (2019g) *Department of Meteorology*. Available at: www.disastermin.gov.lk/web/index.php?option=com_content&view=article&id=53&Itemid=77&lang=en (accessed 2 October 2019).

Ministry of Disaster Management (MDM). (2019h) *Organizational Structure*. Available at: www.disastermin.gov.lk/web/index.php?option=com_content&view=article&id=85&Itemid=146&lang=en (accessed 4 October 2019).

Ministry of Disaster Management and Human Rights (MDM & HR). (2016) *Towards a Safer Sri Lanka: Road Map for Disaster Risk Management: Volume 2 Project Proposals*. Colombo: DMC and MDM&HR.

Ministry of Environment, Climate Change Secretariat. (2010) *National Climate Change Adaptation Strategy for Sri Lanka 2011-2016*. Colombo: Ministry of Environment.

Ministry of National Policy and Economic Affairs and Ministry of Disaster Management (MNPEA & MDM). (2017) Sri Lanka Rapid Post Disaster Needs Assessment Floods and Landslides, May 2017. Colombo: MDM & MNPEA.

Ministry of National Policies and Economic Affairs and Ministry of Disaster Management (MNPEA & MDM). (2016) *Sri Lanka Post Disaster Needs Assessment: Floods and Landslides May 2016*. Colombo: Ministry of Disaster Management.

Peiris, G.H. and Arasaratnam, S. (2019) *Encyclopaedia Britannica – Sri Lanka*. Encyclopaedia Britannica. Available at: www.britannica.com/place/Sri-Lanka (accessed 2 October 2019).

Ratnayake, U. (2013) 'Sea Level Rise' in Disaster Management Centre & Ministry of Disaster Management. (eds.) *Hazard Profiles of Sri Lanka*. Colombo: Disaster Management Centre and Ministry of Disaster Management, pp. 124–139.

Senaratne, P.C. (2013) 'Floods' in Disaster Management Centre & Ministry of Disaster Management. (eds.) *Hazard Profiles of Sri Lanka*. Colombo: Disaster Management Centre and Ministry of Disaster Management, pp. 064–089.

United Nations Development Programme (UNDP). (2016) *Human Development Report 2016*. New York: UNDP.

United Nations Educational, Scientific and Cultural Organization. (2017) *Disaster Risk Reduction: Landslides*. Available at: www.unesco.org/new/en/natural-sciences/special-themes/disaster-risk-reduction/natural-hazards/landslides/ (accessed 2 October 2019).

United Nations Human Settlements Programme (UN-Habitat). (2013) *Cities and Climate Change Initiative Negombo, Sri Lanka: Climate Change Vulnerability Assessment*. Nairobi, Kenya. https://mirror.unhabitat.org/downloads/docs/11942_1_594780.pdf

United Nations World Food Programme. (2017) *National Strategic Review of Food Security and Nutrition 2017*. Colombo: United Nations.

Weerasinghe, K. and Jayatissa, H.A.G. (2013) 'Landslides' in Disaster Management Centre & Ministry of Disaster Management. (eds.) *Hazard Profiles of Sri Lanka*. Colombo: Disaster Management Centre and Ministry of Disaster Management, pp. 090–103.

Wickramaarachchi, B. (2013) 'Coastal Erosion' in Disaster Management Centre & Ministry of Disaster Management. (eds.) *Hazard Profiles of Sri Lanka*. Colombo: Disaster Management Centre and Ministry of Disaster Management, pp. 025–045.

Wijemannage, A.L.K. (2013) 'Tropical Cyclones' in Disaster Management Centre & Ministry of Disaster Management. (eds.) *Hazard Profiles of Sri Lanka*. Colombo: Disaster Management Centre and Ministry of Disaster Management, pp. 174–197.

Wijethunga, J. (2013a) 'Storm Surge' in Disaster Management Centre & Ministry of Disaster Management. (eds.) *Hazard Profiles of Sri Lanka*. Colombo: Disaster Management Centre and Ministry of Disaster Management, pp. 140–173.

Wijethunga, J. (2013b) 'Tsunami' in Disaster Management Centre & Ministry of Disaster Management. (eds.) *Hazard Profiles of Sri Lanka*. Colombo: Disaster Management Centre and Ministry of Disaster Management, pp. 198–231.

World Bank. (2019). *The World Bank in Sri Lanka*. Available at: www.worldbank.org/en/country/srilanka/overview (accessed 3 October 2019).

World Population Review. (2019) *Colombo Population*. Available at: http://worldpopulationreview.com/world-cities/colombo-population/ (accessed 3 October 2019).

Zubair, L., Ralapanawe, V., Tennakoon, U., Yahiya, Z. and Perera, R. (2006) 'Natural Disaster Risks in Sri Lanka: Mapping Hazards and Risk Hotspots' in Arnold, M., Chen, R., Deichmann, U., Dilley, M., Lerner-Lam, A., Pullen, R. and Trohanis, Z. (eds.) *Natural Disaster Hotspots: Case Studies*. Washington, DC: The International Bank for Reconstruction and Development/The World Bank, pp. 109–136.

7 Conclusion

Key considerations for disaster resilience in the smaller countries of South Asia

Overview

The devastating and high-profile disasters that assault the smaller nations of South Asia continue to raise awareness of the importance of building capacity and reducing vulnerability in these nations. The intent of this book was to explore disaster resilience across five of the smaller South Asian countries: Bangladesh, Bhutan, the Maldives, Nepal and Sri Lanka. Although frequently exposed to devastating hazard events, research across many of these regions remains limited; yet is vital to understand vulnerability and build resilience for future events.

The case study countries presented provide a unique insight, given the use of multiple lenses to study the concept of resilience. In particular, the use of a social lens – within the context of a national social-structural backdrop – applied in a community engagement context at the case study level assists with providing a unique opportunity to explore the concept in a holistic way. Importantly, this methodology highlights the different barriers and challenges that communities face in their quest to develop disaster resilience. Understanding options for cultivating resilience within these smaller and often highly vulnerable communities is considered vital, as key influences are highlighted to reinforce policy intent, capacity building and the achievement of disaster resilience on a regional level.

Highlights of the case studies

Bangladesh

Bangladesh is continually subjected to a range of hazards, many of a hydrometeorological nature and is considered one of the most disaster-prone countries globally. Unplanned urbanisation in both major and smaller cities, along with unregulated urban development, potentially expose communities to the catastrophic impact from both natural and human-induced hazards. Recurrent floods and cyclones are major contributors to the ongoing human and economic tolls. Structural and non-structural interventions of government – in the form of a range of policies and processes including built infrastructure initiatives – have had a positive impact, with a reduction in the loss of life following disasters; despite this, economic losses continue to rise. Bangladesh continues to face a high level of poverty, despite recent economic growth – this has been partly aggravated by frequent and recurrent disasters that are increasingly amplified by climate change, highlighting a need to strengthen resilience in a different form.

Bangladesh has been at the forefront of disaster resilience initiatives for a long time, due to recurrent hazards elevating the disaster risk profile and contributing to a significant human toll in the past. The case study, focusing on a project under the Comprehensive Disaster Management Programme (CDMP), identified significant benefits on a local level. These benefits include: the gradual strengthening of policy instruments; initiatives such as CDMP, which since the early 2000s have spearheaded a 'paradigm shift', to from post-disaster response to disaster resilience; and well-funded and carefully developed programmes that have profoundly impacted the thinking across different stakeholders in Bangladesh about how to deal with disasters.

However, the case study revealed major challenges after the completion of the programme. The minimal follow-up strategy of projects has a negative impact on sustainability of outcomes of DRR investments, at least for the government's disaster resilience system. One of the key issues is the instrumental reliance on external funds, influencing governmental decision-making processes, diminishing any initiative for independent action and sustaining investments. Although Bangladesh has been a leader in the disaster resilience field, elevating of Bangladesh to the next level requires a more proactive and innovative approach, not extensively relying on external funding to drive institutional initiatives.

Bhutan

Bhutan is facing a unique disaster resilience challenge due to climate change-driven melting of glaciers causing glacial lake outburst floods (GLOFs). The 1994 GLOF event demonstrated the devastating impact of the disaster and the long-term consequences of such events. The hilly terrain of this country is subjected to rapid and unplanned urbanisation that increases the risk to local communities from a GLOF. Thus, the Bhutan government and its partners rightly prioritised action to reduce the GLOF risk as part of its initial strategies to adapt to climate change. The proactive government initiative NAPA 1, in collaboration with the international community, focused on one of the 22 glacial lakes in the country assessed as posing the most risk, however, reducing disaster risk faced by about 10 per cent of the population, while the remaining 90 per cent is still facing significant disaster risk. NAPA 1 was a start to the continuing and long-term commitment of the Bhutanese government and the international development community to undertake integrated climate change adaptation and disaster resilience measures through a combination of structural and non-structural interventions.

The case study project under NAPA 1 took a multi-pronged approach, including carrying out structural work to lower the water level in the threatening glacial lake, installing an automated EWS in the valleys at risk of GLOFs and building capacity for disaster risk reduction at the institutional and community level. The success of this project – notwithstanding the challenges and its relevance beyond Bhutan for the entire Himalayan-Hindu-Kush mountainous region – resulted in replication in other countries such as Nepal and Pakistan. However, with the predicted rapidly advancing climate change impacts and the ongoing melting of glaciers in the region, the project's efforts of preventing these 'icy floods' in the long run is uncertain. Although this project alone cannot address the mammoth problem in its entirety, it is a concerted effort to do so, thus this project is a hopeful sign for building disaster resilience for complex transboundary issues.

Maldives

The Maldives faces a set of challenges due to ongoing and impending impacts of climate change – including the possibility that parts of the country may go underwater. Moreover, the high-density urbanisation and associated transport and waste management also pose further challenges. Prior to the 2004 tsunami, the Maldives was perceived as a low-hazard country. To some extent that perception is still held, despite the potentially severe impacts of climate change-induced disasters. This has resulted in low budgetary allocation for disaster resilience from national level sources; thus, the nationally mandated governmental body (NDMC) responsible for dealing with disasters is seeking international support for building disaster resilience.

The 2004 Indian Ocean tsunami did not inundate the Maldives completely, yet the on-rush of water flowing over the small, dispersed and low-lying atolls caused significant damage to property and the economy. The Maldives received comparatively less attention from international media and humanitarian agencies, due to relative low impact compared to other countries. The Maldives was primarily serviced by the International Federation of Red Cross (IFRC) and its Red Cross/Crescent national society partners, particularly the American, Australian and British Red Cross, which undertook the bulk of the recovery work.

The post-tsunami recovery project in Dhuvaafaru, a partnership between National Disaster Management Centre and IFRC, fostered many positive aspects including a 'build back better' initiative that has reduced the disaster risk of a resettled community. However, a new set of predicaments have now emerged, such as the uncertain status of property ownership. Despite the disaster resilience efforts incorporated into the project – which earned it the reputation of being a 'safe island' – the protection offered from the imminent threat posed by climate change is limited. Together with the other inhabitants of the Maldives, the Dhuvaafaru residents will need a concerted effort to stay afloat in the face of the rising sea level caused by climate change.

Nepal

Nepal is exposed to a number of natural hazards including earthquakes, floods and landslides, and these hazards have negative impacts upon communities and national economic development. As one of the poorer developing countries, Nepal faces significant challenges including political instabilities, shadows of the past insurgency, significant income inequalities and weak institutions, along with geography-based constraints due to its mountainous topography. Nepal was under-prepared for the 2015 Gorkha earthquake disaster: the impact of the earthquake was overwhelming to the nation as a whole. Nepal had significant international help immediately after the earthquake for relief and reconstruction. However, with an unstable political environment and often underprepared government agencies, there were delays and inefficiencies in the distribution of both the short-term and long-term assistance from the international community. In particular, the political environment and enactment of a new constitution impacted on the execution of the reconstruction phase. Despite these setbacks, the Nepal Parliament passed a new constitution and established the National Reconstruction Authority to oversee the reconstruction process.

Prior to the 2015 earthquake, Nepal had reinforced the role of building codes in developing disaster resilient communities. Since the early 1990s, Nepal continually developed a set of National Building Codes (NBC) and regulations set under several categories with the view of covering all demographics and building types. The Mandatory Rules of Thumb provides a flexible approach, recognising the limitations of low-income and rural communities in adhering to stringent codes. However, many challenges to implementing building codes are posed by Nepal's large rural population, who live in different geological locations with significant income disparity and inadequate presence of government agencies. However, in the post-2015 earthquake environment, a number of new institutions were developed to strengthen the governance of disaster resilience initiatives. Extensive re-training programmes for masons and engineering professionals increased the awareness of building codes. The newly approved 'Disaster Risk Reduction and Management Act 2017' consolidated and replaced many old acts.

In Nepal, many issues impacted the adoption of building codes. Lack of available professional engineering and trades skill in the market was a significant issue as was the 'skills drain' of trained masons leaving the country for better work opportunities, causing constant national skill shortages. The lack of awareness among some professionals of the importance of NBC and its role in disaster resilience also contributes to Nepal's vulnerability to earthquakes. This can be addressed, for example, by continuous training and improvements, and also by addressing the gender balance in masonry skills. Inadequate institutional capacity posed a further barrier to effective building code implementation. Despite the improvements to government institutions after the 2015 earthquake, inadequate institutional capacity in terms of skilled technical human resources and scarcity of financial resources impedes the efficiency of the government.

Sri Lanka

Sri Lanka experiences a range of different disaster events from storms and floods through to drought conditions. The south-west and north-east monsoon seasons often led to heavy rains and subsequent flooding and landslides triggered by inundation. The high degree of impacts experienced on a local level have been exacerbated by rapid urbanisation and widespread unauthorised development activities. Since the 2004 Indian Ocean Tsunami, the government has been proactive in disaster resilience activities; however, the 2017 flooding and landslide events exposed local communities to severe devastation with significant loss of life, damage to property and infrastructure, identifying another level of vulnerability.

The 2017 landslides in Ratnapura district reinforced the challenges experienced at the local level. In response to the event, a need to strengthen policy and increase resources has been identified – particularly in relation to organisational coordination and communication, early warning systems, and planning and building enforcement. Issues that contributed to the devastating impact include the lack of an appropriate hazard forecasting technology to enhance early warning activities that caters to a wider region, rather than concentrating on small areas, and the lack of community response, for example, ignoring media communications. Behavioural issues such as unwillingness of some community members to cooperate with the authorities and evacuate made the situation more difficult.

Lacking an overriding coordinating body to manage disasters, along with multiple disparate agencies supporting relief provision and other emergency management activities meant that relief provision was often duplicated, resources wasted, and delays experienced in responding with needed services. Furthermore, the lack of emergency response and relief operations, difficulties with essential supplies, equipment and lack of coordination at the local level elevated the impact. Increased community preparedness and institutional flexibility are fundamental to building resilience. Nonetheless, positive change is emerging as formalised shelters are established, education of masons is being undertaken to 'build back better', and policy is being enacted to manage rescue and relief operations. Learning from these activities and other areas in need of consideration will assist local communities to achieve a level of resilience to minimise future devastation.

Key observations and lessons for the future

Building disaster resilience is a complex process involving a multitude of stakeholders, institutions and complex socio-cultural political boundaries that need to be contemplated. Disaster resilience is therefore the business of governments, but extends beyond that distinct sphere to involve every part of society: public and private entities, regional and local-level actors. Across all five case studies, country-specific issues were identified that need to be addressed if resilience is to be achieved; however, there were five main considerations present across all cases and these will now be discussed.

Climate change and rapid urbanisation placing pressure on disaster resilience efforts

The impact of climate change is clearly evident in the patterns of increasing frequency and impact of natural hazards. Moreover, economic development pressures leading to rapid urbanisation and non-compliant infrastructure is placing pressure on already vulnerable systems. Changes to land through urban development activities are widespread, and significantly impact the natural environment and its ability to cope with disaster events. Deforestation, land clearance for agriculture, as well as illegal developments and waste management practices all contribute to modification of land and subsequently increase the risk of disasters. Although there is no easy remedy, planning and building policy has the potential to minimise risk through, for example, conducting development outside of high-risk zones. Therefore, hazard risk mapping provides an alternative as it clearly designates suitable development areas to minimise risk and exposure. However, resources and education are mandatory for the success of these types of initiatives and as such rely upon the commitment of government and the private sector for support.

Disaster resilience initiatives are supported by external funding with limited follow-up afterwards

Most South Asian nations were and continue to be heavily supported by international funding and international non-governmental organisations (NGOs), as they have limited capabilities to prepare for, respond to and reconstruct after disasters. The conditions placed on these nations by the international funding bodies in the past have to some extent emphasised developing better institutions in these countries – particularly government agencies – to manage funds and implement projects and programmes effectively. Some of these institutions have been made permanent, beyond the funding period, and are positive developments that have improved the disaster resilience capacities of these countries. Nonetheless, local institutional initiatives that stem from these capacities would need to define future disaster resilience.

Although internal funding and engagement provide much-needed support to these nations, there are also some negative outcomes evident. Expectations of international funding to develop local capacities or even identify local disaster resilience initiatives make these countries less proactive and dependent on international intervention. Moreover, the lack of national level resourcing for running these institutions on a regular basis is making many of these institutions' ongoing contributions limited. The governments are more prone to be only interested in large structural or physical infrastructure interventions that need a high level of funding, and some critics associate that with potential for corruption.

There is a culture of lack of follow-up or maintenance of the initiatives arising from the international funding. The capacities or physical infrastructure developed using international funding are mostly not maintained or further developed, which in some cases increases vulnerability and reduces the resilience of communities over a period of time. Often a lack of national level support and resourcing is blamed for this predicament.

Inter-organisational coordination/collaboration and ongoing capacity development

Ineffective or lack of communication and coordination internally and externally, and between government agencies and NGOs, has resulted in duplication of services and ineffective outcomes to communities, such as lack of provision of services during disaster response. In some cases, the lack of coordination has led to increased vulnerability because agencies did not identify imminent hazards. Political stability is an important aspect for building institutions that can develop disaster mitigation, response and reconstruction activities. In most of the case study countries, political instability created significant delays in the response and reconstruction process, and in the overall disaster resilience process.

Although in almost all of the five countries there were well-established institutions, regulations and legislations, they needed to address bottlenecks in terms of:

- Bureaucratic processes delaying project implementation, relief and rescue operations.
- Officials not wishing to work outside of standard operating procedures, limiting innovative approaches to context-specific problems.
- Residents not responding to early warnings and wishing to evacuate or delaying evacuation until impossible.
- Gathering required information, especially individual and community information, that helps effective disaster risk reduction and management.
- Tension between formal and informal processes that had divergent objectives.
- Achieving community consultation and engagement for disaster resilience initiatives.

Typically, this concerns areas like forecasting and early warning systems, and collaboration amongst key stakeholders – both government and non-government – in conjunction with operational planning. Moreover, government agencies need to upgrade with more advanced technology to assist with forecasting, early warning and risk/hazard mapping. Ongoing investment in capturing and updating knowledge and skills is key for disaster resilience.

Proactive policy development and information platforms for achieving disaster resilience

In many South Asian countries, particularly in rural areas, land ownership details are not maintained at the government level. Moreover, lack of government records on migration, birth and address of citizens creates massive problems for distributing disaster relief and reconstruction support. Many disaster survivors who were living in unregistered property without land ownership and title had significant challenges in proving their entitlement to support. Relying on local knowledge and social setup to make decisions on resource distribution in the absence of government recourse leads to nepotism, marginalisation of vulnerable communities and differentiated treatment based on social power, eventually creating more vulnerability among communities. Therefore, governments should develop fair and transparent protocols for identifying the eligibility for disaster relief, with room to fine-tune based on the level of devastation or complexity of actual events.

Appropriate policy is vital to build resilience on a local level. Policy encompasses areas such as planning and building as mentioned previously, however, it also applies to preparedness, and disaster risk reduction and management processes. Importantly, policy must be accompanied with education and awareness. It is vital to have policy in its various forms, but without compliance it remains just a document. Policy compliance is generally thought of as being aligned with enforcement – this is often fraught with issues, given a lack of sufficient resources to support such activities, particularly within vulnerable and underprivileged communities. However, in many cases education has been a mechanism by which to encourage voluntary compliance, particularly in relation to planning and building regulations, and may offer an opportunity in the context of disaster resilience.

Market development and education relating to the constructing sector

Government agencies, NGOs and the construction sector – as either commercial entities or social enterprises – play a significant role in building disaster resilience. However, owner-driven construction is a significant proportion of the residential construction sector in developing countries, particularly in rural areas. Low quality and non-compliant construction is common in these countries, increasing disaster risk. Governments have engaged in substantial re-education initiatives to improve the knowledge and skills of the sector relating to compliant construction. The massive re-training programmes for masons and engineers is an example of such education and market development. However, labour movement and skills drain are creating labour and skill shortages. In reconstruction projects, new housing is often imbued with a level of safe and good-quality construction, but subsequent transformations by beneficiaries usher in disaster risk – in such cases, long-term training and a market sector that is responsive to the need for future resilience is essential.

Concluding remarks

The ability to achieve absolute resilience is an intangible concept, if not unimaginable. Yet through actions, *inter alia*, collaboration, awareness, commitment, transparency and importantly acceptance, there is potential to build capacity and minimise the devastation that continues to assault these nations. The opportunity to move forward and create resilient communities relies, therefore, on a multitude of key principles, a number of which have been presented in this book. These endeavours are but one small part in the cycle of learning and change that may assist to achieve positive outcomes through knowledge dissemination and identifying challenges as well as barriers and enablers to change.

Index

Page numbers in *italics* refer to images; those in **bold** indicate tables.

For Product Safety Concerns and Information please contact our EU representative GPSR@taylorandfrancis.com
Taylor & Francis Verlag GmbH, Kaufingerstraße 24, 80331 München, Germany

www.ingramcontent.com/pod-product-compliance
Ingram Content Group UK Ltd.
Pitfield, Milton Keynes, MK11 3LW, UK
UKHW021302190926
475701UK00009B/579
9781032236360